Planets: A Very Short Introduction

VERY SHORT INTRODUCTIONS are for anyone wanting a stimulating and accessible way into a new subject. They are written by experts, and have been translated into more than 45 different languages.

The series began in 1995, and now covers a wide variety of topics in every discipline. The VSI library currently contains over 650 volumes—a Very Short Introduction to everything from Psychology and Philosophy of Science to American History and Relativity—and continues to grow in every subject area.

Very Short Introductions available now:

ABOLITIONISM Richard S. Newman
THE ABRAHAMIC RELIGIONS Charles L. Cohen
ACCOUNTING Christopher Nobes
ADOLESCENCE Peter K. Smith
ADVERTISING Winston Fletcher
AERIAL WARFARE Frank Ledwidge
AESTHETICS Bence Nanay
AFRICAN AMERICAN RELIGION Eddie S. Glaude Jr
AFRICAN HISTORY John Parker and Richard Rathbone
AFRICAN POLITICS Ian Taylor
AFRICAN RELIGIONS Jacob K. Olupona
AGEING Nancy A. Pachana
AGNOSTICISM Robin Le Poidevin
AGRICULTURE Paul Brassley and Richard Soffe
ALEXANDER THE GREAT Hugh Bowden
ALGEBRA Peter M. Higgins
AMERICAN BUSINESS HISTORY Walter A. Friedman
AMERICAN CULTURAL HISTORY Eric Avila
AMERICAN FOREIGN RELATIONS Andrew Preston
AMERICAN HISTORY Paul S. Boyer
AMERICAN IMMIGRATION David A. Gerber
AMERICAN LEGAL HISTORY G. Edward White
AMERICAN MILITARY HISTORY Joseph T. Glatthaar
AMERICAN NAVAL HISTORY Craig L. Symonds
AMERICAN POLITICAL HISTORY Donald Critchlow
AMERICAN POLITICAL PARTIES AND ELECTIONS L. Sandy Maisel
AMERICAN POLITICS Richard M. Valelly
THE AMERICAN PRESIDENCY Charles O. Jones
THE AMERICAN REVOLUTION Robert J. Allison
AMERICAN SLAVERY Heather Andrea Williams
THE AMERICAN SOUTH Charles Reagan Wilson
THE AMERICAN WEST Stephen Aron
AMERICAN WOMEN'S HISTORY Susan Ware
AMPHIBIANS T. S. Kemp
ANAESTHESIA Aidan O'Donnell
ANALYTIC PHILOSOPHY Michael Beaney
ANARCHISM Colin Ward
ANCIENT ASSYRIA Karen Radner
ANCIENT EGYPT Ian Shaw
ANCIENT EGYPTIAN ART AND ARCHITECTURE Christina Riggs
ANCIENT GREECE Paul Cartledge
THE ANCIENT NEAR EAST Amanda H. Podany
ANCIENT PHILOSOPHY Julia Annas
ANCIENT WARFARE Harry Sidebottom
ANGELS David Albert Jones

ANGLICANISM Mark Chapman
THE ANGLO-SAXON AGE John Blair
ANIMAL BEHAVIOUR
Tristram D. Wyatt
THE ANIMAL KINGDOM
Peter Holland
ANIMAL RIGHTS David DeGrazia
THE ANTARCTIC Klaus Dodds
ANTHROPOCENE Erle C. Ellis
ANTISEMITISM Steven Beller
ANXIETY Daniel Freeman and
Jason Freeman
THE APOCRYPHAL GOSPELS
Paul Foster
APPLIED MATHEMATICS
Alain Goriely
THOMAS AQUINAS Fergus Kerr
ARBITRATION Thomas Schultz and
Thomas Grant
ARCHAEOLOGY Paul Bahn
ARCHITECTURE Andrew Ballantyne
ARISTOCRACY William Doyle
ARISTOTLE Jonathan Barnes
ART HISTORY Dana Arnold
ART THEORY Cynthia Freeland
ARTIFICIAL INTELLIGENCE
Margaret A. Boden
ASIAN AMERICAN HISTORY
Madeline Y. Hsu
ASTROBIOLOGY David C. Catling
ASTROPHYSICS James Binney
ATHEISM Julian Baggini
THE ATMOSPHERE Paul I. Palmer
AUGUSTINE Henry Chadwick
AUSTRALIA Kenneth Morgan
AUTISM Uta Frith
AUTOBIOGRAPHY Laura Marcus
THE AVANT GARDE David Cottington
THE AZTECS Davíd Carrasco
BABYLONIA Trevor Bryce
BACTERIA Sebastian G. B. Amyes
BANKING John Goddard and
John O. S. Wilson
BARTHES Jonathan Culler
THE BEATS David Sterritt
BEAUTY Roger Scruton
BEHAVIOURAL ECONOMICS
Michelle Baddeley
BESTSELLERS John Sutherland
THE BIBLE John Riches
BIBLICAL ARCHAEOLOGY
Eric H. Cline
BIG DATA Dawn E. Holmes
BIOCHEMISTRY Mark Lorch
BIOGEOGRAPHY Mark V. Lomolino
BIOGRAPHY Hermione Lee
BIOMETRICS Michael Fairhurst
BLACK HOLES Katherine Blundell
BLASPHEMY Yvonne Sherwood
BLOOD Chris Cooper
THE BLUES Elijah Wald
THE BODY Chris Shilling
THE BOOK OF COMMON PRAYER
Brian Cummings
THE BOOK OF MORMON
Terryl Givens
BORDERS Alexander C. Diener and
Joshua Hagen
THE BRAIN Michael O'Shea
BRANDING Robert Jones
THE BRICS Andrew F. Cooper
THE BRITISH CONSTITUTION
Martin Loughlin
THE BRITISH EMPIRE Ashley Jackson
BRITISH POLITICS Tony Wright
BUDDHA Michael Carrithers
BUDDHISM Damien Keown
BUDDHIST ETHICS Damien Keown
BYZANTIUM Peter Sarris
CALVINISM Jon Balserak
ALBERT CAMUS Oliver Gloag
CANADA Donald Wright
CANCER Nicholas James
CAPITALISM James Fulcher
CATHOLICISM Gerald O'Collins
CAUSATION Stephen Mumford and
Rani Lill Anjum
THE CELL Terence Allen and
Graham Cowling
THE CELTS Barry Cunliffe
CHAOS Leonard Smith
GEOFFREY CHAUCER David Wallace
CHEMISTRY Peter Atkins
CHILD PSYCHOLOGY Usha Goswami
CHILDREN'S LITERATURE
Kimberley Reynolds
CHINESE LITERATURE Sabina Knight
CHOICE THEORY Michael Allingham
CHRISTIAN ART Beth Williamson
CHRISTIAN ETHICS D. Stephen Long

CHRISTIANITY Linda Woodhead
CIRCADIAN RHYTHMS Russell Foster and Leon Kreitzman
CITIZENSHIP Richard Bellamy
CITY PLANNING Carl Abbott
CIVIL ENGINEERING David Muir Wood
CLASSICAL LITERATURE William Allan
CLASSICAL MYTHOLOGY Helen Morales
CLASSICS Mary Beard and John Henderson
CLAUSEWITZ Michael Howard
CLIMATE Mark Maslin
CLIMATE CHANGE Mark Maslin
CLINICAL PSYCHOLOGY Susan Llewelyn and Katie Aafjes-van Doorn
COGNITIVE NEUROSCIENCE Richard Passingham
THE COLD WAR Robert J. McMahon
COLONIAL AMERICA Alan Taylor
COLONIAL LATIN AMERICAN LITERATURE Rolena Adorno
COMBINATORICS Robin Wilson
COMEDY Matthew Bevis
COMMUNISM Leslie Holmes
COMPARATIVE LITERATURE Ben Hutchinson
COMPETITION AND ANTITRUST LAW Ariel Ezrachi
COMPLEXITY John H. Holland
THE COMPUTER Darrel Ince
COMPUTER SCIENCE Subrata Dasgupta
CONCENTRATION CAMPS Dan Stone
CONFUCIANISM Daniel K. Gardner
THE CONQUISTADORS Matthew Restall and Felipe Fernández-Armesto
CONSCIENCE Paul Strohm
CONSCIOUSNESS Susan Blackmore
CONTEMPORARY ART Julian Stallabrass
CONTEMPORARY FICTION Robert Eaglestone
CONTINENTAL PHILOSOPHY Simon Critchley
COPERNICUS Owen Gingerich
CORAL REEFS Charles Sheppard
CORPORATE SOCIAL RESPONSIBILITY Jeremy Moon
CORRUPTION Leslie Holmes
COSMOLOGY Peter Coles
COUNTRY MUSIC Richard Carlin
CREATIVITY Vlad Glăveanu
CRIME FICTION Richard Bradford
CRIMINAL JUSTICE Julian V. Roberts
CRIMINOLOGY Tim Newburn
CRITICAL THEORY Stephen Eric Bronner
THE CRUSADES Christopher Tyerman
CRYPTOGRAPHY Fred Piper and Sean Murphy
CRYSTALLOGRAPHY A. M. Glazer
THE CULTURAL REVOLUTION Richard Curt Kraus
DADA AND SURREALISM David Hopkins
DANTE Peter Hainsworth and David Robey
DARWIN Jonathan Howard
THE DEAD SEA SCROLLS Timothy H. Lim
DECADENCE David Weir
DECOLONIZATION Dane Kennedy
DEMENTIA Kathleen Taylor
DEMOCRACY Bernard Crick
DEMOGRAPHY Sarah Harper
DEPRESSION Jan Scott and Mary Jane Tacchi
DERRIDA Simon Glendinning
DESCARTES Tom Sorell
DESERTS Nick Middleton
DESIGN John Heskett
DEVELOPMENT Ian Goldin
DEVELOPMENTAL BIOLOGY Lewis Wolpert
THE DEVIL Darren Oldridge
DIASPORA Kevin Kenny
CHARLES DICKENS Jenny Hartley
DICTIONARIES Lynda Mugglestone
DINOSAURS David Norman
DIPLOMACATIC HISTORY Joseph M. Siracusa
DOCUMENTARY FILM Patricia Aufderheide
DREAMING J. Allan Hobson
DRUGS Les Iversen
DRUIDS Barry Cunliffe

DYNASTY Jeroen Duindam
DYSLEXIA Margaret J. Snowling
EARLY MUSIC Thomas Forrest Kelly
THE EARTH Martin Redfern
EARTH SYSTEM SCIENCE Tim Lenton
ECOLOGY Jaboury Ghazoul
ECONOMICS Partha Dasgupta
EDUCATION Gary Thomas
EGYPTIAN MYTH Geraldine Pinch
EIGHTEENTH-CENTURY BRITAIN Paul Langford
THE ELEMENTS Philip Ball
EMOTION Dylan Evans
EMPIRE Stephen Howe
ENERGY SYSTEMS Nick Jenkins
ENGELS Terrell Carver
ENGINEERING David Blockley
THE ENGLISH LANGUAGE Simon Horobin
ENGLISH LITERATURE Jonathan Bate
THE ENLIGHTENMENT John Robertson
ENTREPRENEURSHIP Paul Westhead and Mike Wright
ENVIRONMENTAL ECONOMICS Stephen Smith
ENVIRONMENTAL ETHICS Robin Attfield
ENVIRONMENTAL LAW Elizabeth Fisher
ENVIRONMENTAL POLITICS Andrew Dobson
ENZYMES Paul Engel
EPICUREANISM Catherine Wilson
EPIDEMIOLOGY Rodolfo Saracci
ETHICS Simon Blackburn
ETHNOMUSICOLOGY Timothy Rice
THE ETRUSCANS Christopher Smith
EUGENICS Philippa Levine
THE EUROPEAN UNION Simon Usherwood and John Pinder
EUROPEAN UNION LAW Anthony Arnull
EVOLUTION Brian and Deborah Charlesworth
EXISTENTIALISM Thomas Flynn
EXPLORATION Stewart A. Weaver
EXTINCTION Paul B. Wignall
THE EYE Michael Land
FAIRY TALE Marina Warner
FAMILY LAW Jonathan Herring
MICHAEL FARADAY Frank A. J. L. James
FASCISM Kevin Passmore
FASHION Rebecca Arnold
FEDERALISM Mark J. Rozell and Clyde Wilcox
FEMINISM Margaret Walters
FILM Michael Wood
FILM MUSIC Kathryn Kalinak
FILM NOIR James Naremore
FIRE Andrew C. Scott
THE FIRST WORLD WAR Michael Howard
FOLK MUSIC Mark Slobin
FOOD John Krebs
FORENSIC PSYCHOLOGY David Canter
FORENSIC SCIENCE Jim Fraser
FORESTS Jaboury Ghazoul
FOSSILS Keith Thomson
FOUCAULT Gary Gutting
THE FOUNDING FATHERS R. B. Bernstein
FRACTALS Kenneth Falconer
FREE SPEECH Nigel Warburton
FREE WILL Thomas Pink
FREEMASONRY Andreas Önnerfors
FRENCH LITERATURE John D. Lyons
FRENCH PHILOSOPHY Stephen Gaukroger and Knox Peden
THE FRENCH REVOLUTION William Doyle
FREUD Anthony Storr
FUNDAMENTALISM Malise Ruthven
FUNGI Nicholas P. Money
THE FUTURE Jennifer M. Gidley
GALAXIES John Gribbin
GALILEO Stillman Drake
GAME THEORY Ken Binmore
GANDHI Bhikhu Parekh
GARDEN HISTORY Gordon Campbell
GENES Jonathan Slack
GENIUS Andrew Robinson
GENOMICS John Archibald
GEOGRAPHY John Matthews and David Herbert
GEOLOGY Jan Zalasiewicz
GEOPHYSICS William Lowrie

GEOPOLITICS Klaus Dodds
GERMAN LITERATURE Nicholas Boyle
GERMAN PHILOSOPHY Andrew Bowie
THE GHETTO Bryan Cheyette
GLACIATION David J. A. Evans
GLOBAL CATASTROPHES Bill McGuire
GLOBAL ECONOMIC HISTORY Robert C. Allen
GLOBAL ISLAM Nile Green
GLOBALIZATION Manfred B. Steger
GOD John Bowker
GOETHE Ritchie Robertson
THE GOTHIC Nick Groom
GOVERNANCE Mark Bevir
GRAVITY Timothy Clifton
THE GREAT DEPRESSION AND THE NEW DEAL Eric Rauchway
HABEAS CORPUS Amanda Tyler
HABERMAS James Gordon Finlayson
THE HABSBURG EMPIRE Martyn Rady
HAPPINESS Daniel M. Haybron
THE HARLEM RENAISSANCE Cheryl A. Wall
THE HEBREW BIBLE AS LITERATURE Tod Linafelt
HEGEL Peter Singer
HEIDEGGER Michael Inwood
THE HELLENISTIC AGE Peter Thonemann
HEREDITY John Waller
HERMENEUTICS Jens Zimmermann
HERODOTUS Jennifer T. Roberts
HIEROGLYPHS Penelope Wilson
HINDUISM Kim Knott
HISTORY John H. Arnold
THE HISTORY OF ASTRONOMY Michael Hoskin
THE HISTORY OF CHEMISTRY William H. Brock
THE HISTORY OF CHILDHOOD James Marten
THE HISTORY OF CINEMA Geoffrey Nowell-Smith
THE HISTORY OF LIFE Michael Benton
THE HISTORY OF MATHEMATICS Jacqueline Stedall
THE HISTORY OF MEDICINE William Bynum
THE HISTORY OF PHYSICS J. L. Heilbron
THE HISTORY OF TIME Leofranc Holford-Strevens
HIV AND AIDS Alan Whiteside
HOBBES Richard Tuck
HOLLYWOOD Peter Decherney
THE HOLY ROMAN EMPIRE Joachim Whaley
HOME Michael Allen Fox
HOMER Barbara Graziosi
HORMONES Martin Luck
HORROR Darryl Jones
HUMAN ANATOMY Leslie Klenerman
HUMAN EVOLUTION Bernard Wood
HUMAN PHYSIOLOGY Jamie A. Davies
HUMAN RIGHTS Andrew Clapham
HUMANISM Stephen Law
HUME A. J. Ayer
HUMOUR Noël Carroll
THE ICE AGE Jamie Woodward
IDENTITY Florian Coulmas
IDEOLOGY Michael Freeden
THE IMMUNE SYSTEM Paul Klenerman
INDIAN CINEMA Ashish Rajadhyaksha
INDIAN PHILOSOPHY Sue Hamilton
THE INDUSTRIAL REVOLUTION Robert C. Allen
INFECTIOUS DISEASE Marta L. Wayne and Benjamin M. Bolker
INFINITY Ian Stewart
INFORMATION Luciano Floridi
INNOVATION Mark Dodgson and David Gann
INTELLECTUAL PROPERTY Siva Vaidhyanathan
INTELLIGENCE Ian J. Deary
INTERNATIONAL LAW Vaughan Lowe
INTERNATIONAL MIGRATION Khalid Koser
INTERNATIONAL RELATIONS Christian Reus-Smit

INTERNATIONAL SECURITY
Christopher S. Browning
IRAN Ali M. Ansari
ISLAM Malise Ruthven
ISLAMIC HISTORY Adam Silverstein
ISLAMIC LAW Mashood A. Baderin
ISOTOPES Rob Ellam
ITALIAN LITERATURE
Peter Hainsworth and David Robey
HENRY JAMES Susan L. Mizruchi
JESUS Richard Bauckham
JEWISH HISTORY David N. Myers
JOURNALISM Ian Hargreaves
JUDAISM Norman Solomon
JUNG Anthony Stevens
KABBALAH Joseph Dan
KAFKA Ritchie Robertson
KANT Roger Scruton
KEYNES Robert Skidelsky
KIERKEGAARD Patrick Gardiner
KNOWLEDGE Jennifer Nagel
THE KORAN Michael Cook
KOREA Michael J. Seth
LAKES Warwick F. Vincent
LANDSCAPE ARCHITECTURE
Ian H. Thompson
LANDSCAPES AND GEOMORPHOLOGY
Andrew Goudie and Heather Viles
LANGUAGES Stephen R. Anderson
LATE ANTIQUITY Gillian Clark
LAW Raymond Wacks
THE LAWS OF THERMODYNAMICS
Peter Atkins
LEADERSHIP Keith Grint
LEARNING Mark Haselgrove
LEIBNIZ Maria Rosa Antognazza
C. S. LEWIS James Como
LIBERALISM Michael Freeden
LIGHT Ian Walmsley
LINCOLN Allen C. Guelzo
LINGUISTICS Peter Matthews
LITERARY THEORY Jonathan Culler
LOCKE John Dunn
LOGIC Graham Priest
LOVE Ronald de Sousa
MARTIN LUTHER Scott H. Hendrix
MACHIAVELLI Quentin Skinner
MADNESS Andrew Scull
MAGIC Owen Davies
MAGNA CARTA Nicholas Vincent
MAGNETISM Stephen Blundell
MALTHUS Donald Winch
MAMMALS T. S. Kemp
MANAGEMENT John Hendry
NELSON MANDELA Elleke Boehmer
MAO Delia Davin
MARINE BIOLOGY Philip V. Mladenov
MARKETING
Kenneth Le Meunier-FitzHugh
THE MARQUIS DE SADE
John Phillips
MARTYRDOM Jolyon Mitchell
MARX Peter Singer
MATERIALS Christopher Hall
MATHEMATICAL FINANCE Mark H. A. Davis
MATHEMATICS Timothy Gowers
MATTER Geoff Cottrell
THE MAYA Matthew Restall and Amara Solari
THE MEANING OF LIFE
Terry Eagleton
MEASUREMENT David Hand
MEDICAL ETHICS Michael Dunn and Tony Hope
MEDICAL LAW Charles Foster
MEDIEVAL BRITAIN John Gillingham and Ralph A. Griffiths
MEDIEVAL LITERATURE
Elaine Treharne
MEDIEVAL PHILOSOPHY
John Marenbon
MEMORY Jonathan K. Foster
METAPHYSICS Stephen Mumford
METHODISM William J. Abraham
THE MEXICAN REVOLUTION
Alan Knight
MICROBIOLOGY Nicholas P. Money
MICROECONOMICS Avinash Dixit
MICROSCOPY Terence Allen
THE MIDDLE AGES Miri Rubin
MILITARY JUSTICE Eugene R. Fidell
MILITARY STRATEGY
Antulio J. Echevarria II
MINERALS David Vaughan
MIRACLES Yujin Nagasawa
MODERN ARCHITECTURE
Adam Sharr
MODERN ART David Cottington

MODERN BRAZIL Anthony W. Pereira
MODERN CHINA Rana Mitter
MODERN DRAMA
Kirsten E. Shepherd-Barr
MODERN FRANCE
Vanessa R. Schwartz
MODERN INDIA Craig Jeffrey
MODERN IRELAND Senia Pašeta
MODERN ITALY Anna Cento Bull
MODERN JAPAN
Christopher Goto-Jones
MODERN LATIN AMERICAN LITERATURE
Roberto González Echevarría
MODERN WAR Richard English
MODERNISM Christopher Butler
MOLECULAR BIOLOGY Aysha Divan and Janice A. Royds
MOLECULES Philip Ball
MONASTICISM Stephen J. Davis
THE MONGOLS Morris Rossabi
MONTAIGNE William M. Hamlin
MOONS David A. Rothery
MORMONISM RICHARD Lyman Bushman
MOUNTAINS Martin F. Price
MUHAMMAD Jonathan A. C. Brown
MULTICULTURALISM Ali Rattansi
MULTILINGUALISM John C. Maher
MUSIC Nicholas Cook
MYTH Robert A. Segal
NAPOLEON David Bell
THE NAPOLEONIC WARS
Mike Rapport
NATIONALISM Steven Grosby
NATIVE AMERICAN LITERATURE
Sean Teuton
NAVIGATION Jim Bennett
NAZI GERMANY Jane Caplan
NEOLIBERALISM Manfred B. Steger and Ravi K. Roy
NETWORKS Guido Caldarelli and Michele Catanzaro
THE NEW TESTAMENT
Luke Timothy Johnson
THE NEW TESTAMENT AS LITERATURE Kyle Keefer
NEWTON Robert Iliffe
NIELS BOHR J. L. Heilbron
NIETZSCHE Michael Tanner
NINETEENTH-CENTURY BRITAIN
Christopher Harvie and H. C. G. Matthew
THE NORMAN CONQUEST
George Garnett
NORTH AMERICAN INDIANS
Theda Perdue and Michael D. Green
NORTHERN IRELAND
Marc Mulholland
NOTHING Frank Close
NUCLEAR PHYSICS Frank Close
NUCLEAR POWER Maxwell Irvine
NUCLEAR WEAPONS
Joseph M. Siracusa
NUMBER THEORY Robin Wilson
NUMBERS Peter M. Higgins
NUTRITION David A. Bender
OBJECTIVITY Stephen Gaukroger
OCEANS Dorrik Stow
THE OLD TESTAMENT
Michael D. Coogan
THE ORCHESTRA D. Kern Holoman
ORGANIC CHEMISTRY
Graham Patrick
ORGANIZATIONS Mary Jo Hatch
ORGANIZED CRIME
Georgios A. Antonopoulos and Georgios Papanicolaou
ORTHODOX CHRISTIANITY
A. Edward Siecienski
OVID Llewelyn Morgan
PAGANISM Owen Davies
THE PALESTINIAN-ISRAELI CONFLICT Martin Bunton
PANDEMICS Christian W. McMillen
PARTICLE PHYSICS Ffank Close
PAUL E. P. Sanders
PEACE Oliver P. Richmond
PENTECOSTALISM
William K. Kay
PERCEPTION Brian Rogers
THE PERIODIC TABLE
Eric R. Scerri
PHILOSOPHICAL METHOD
Timothy Williamson
PHILOSOPHY Edward Craig
PHILOSOPHY IN THE ISLAMIC WORLD Peter Adamson
PHILOSOPHY OF BIOLOGY
Samir Okasha

PHILOSOPHY OF LAW Raymond Wacks
PHILOSOPHY OF PHYSICS David Wallace
PHILOSOPHY OF SCIENCE Samir Okasha
PHILOSOPHY OF RELIGION Tim Bayne
PHOTOGRAPHY Steve Edwards
PHYSICAL CHEMISTRY Peter Atkins
PHYSICS Sidney Perkowitz
PILGRIMAGE Ian Reader
PLAGUE Paul Slack
PLANETS David A. Rothery
PLANTS Timothy Walker
PLATE TECTONICS Peter Molnar
PLATO Julia Annas
POETRY Bernard O'Donoghue
POLITICAL PHILOSOPHY David Miller
POLITICS Kenneth Minogue
POPULISM Cas Mudde and Cristóbal Rovira Kaltwasser
POSTCOLONIALISM Robert Young
POSTMODERNISM Christopher Butler
POSTSTRUCTURALISM Catherine Belsey
POVERTY Philip N. Jefferson
PREHISTORY Chris Gosden
PRESOCRATIC PHILOSOPHY Catherine Osborne
PRIVACY Raymond Wacks
PROBABILITY John Haigh
PROGRESSIVISM Walter Nugent
PROHIBITION W. J. Rorabaugh
PROJECTS Andrew Davies
PROTESTANTISM Mark A. Noll
PSYCHIATRY Tom Burns
PSYCHOANALYSIS Daniel Pick
PSYCHOLOGY Gillian Butler and Freda McManus
PSYCHOLOGY OF MUSIC Elizabeth Hellmuth Margulis
PSYCHOPATHY Essi Viding
PSYCHOTHERAPY Tom Burns and Eva Burns-Lundgren
PUBLIC ADMINISTRATION Stella Z. Theodoulou and Ravi K. Roy
PUBLIC HEALTH Virginia Berridge
PURITANISM Francis J. Bremer
THE QUAKERS Pink Dandelion
QUANTUM THEORY John Polkinghorne
RACISM Ali Rattansi
RADIOACTIVITY Claudio Tuniz
RASTAFARI Ennis B. Edmonds
READING Belinda Jack
THE REAGAN REVOLUTION Gil Troy
REALITY Jan Westerhoff
RECONSTRUCTION Allen C. Guelzo
THE REFORMATION Peter Marshall
REFUGEES Gil Loescher
RELATIVITY Russell Stannard
RELIGION Thomas A. Tweed
RELIGION IN AMERICA Timothy Beal
THE RENAISSANCE Jerry Brotton
RENAISSANCE ART Geraldine A. Johnson
RENEWABLE ENERGY Nick Jelley
REPTILES T.S. Kemp
REVOLUTIONS Jack A. Goldstone
RHETORIC Richard Toye
RISK Baruch Fischhoff and John Kadvany
RITUAL Barry Stephenson
RIVERS Nick Middleton
ROBOTICS Alan Winfield
ROCKS Jan Zalasiewicz
ROMAN BRITAIN Peter Salway
THE ROMAN EMPIRE Christopher Kelly
THE ROMAN REPUBLIC David M. Gwynn
ROMANTICISM Michael Ferber
ROUSSEAU Robert Wokler
RUSSELL A. C. Grayling
THE RUSSIAN ECONOMY Richard Connolly
RUSSIAN HISTORY Geoffrey Hosking
RUSSIAN LITERATURE Catriona Kelly
THE RUSSIAN REVOLUTION S. A. Smith
THE SAINTS Simon Yarrow
SAMURAI Michael Wert
SAVANNAS Peter A. Furley
SCEPTICISM Duncan Pritchard
SCHIZOPHRENIA Chris Frith and Eve Johnstone
SCHOPENHAUER Christopher Janaway
SCIENCE AND RELIGION Thomas Dixon

SCIENCE FICTION David Seed
THE SCIENTIFIC REVOLUTION Lawrence M. Principe
SCOTLAND Rab Houston
SECULARISM Andrew Copson
SEXUAL SELECTION Marlene Zuk and Leigh W. Simmons
SEXUALITY Véronique Mottier
WILLIAM SHAKESPEARE Stanley Wells
SHAKESPEARE'S COMEDIES Bart van Es
SHAKESPEARE'S SONNETS AND POEMS Jonathan F. S. Post
SHAKESPEARE'S TRAGEDIES Stanley Wells
GEORGE BERNARD SHAW Christopher Wixson
SIKHISM Eleanor Nesbitt
SILENT FILM Donna Kornhaber
THE SILK ROAD James A. Millward
SLANG Jonathon Green
SLEEP Steven W. Lockley and Russell G. Foster
SMELL Matthew Cobb
ADAM SMITH Christopher J. Berry
SOCIAL AND CULTURAL ANTHROPOLOGY John Monaghan and Peter Just
SOCIAL PSYCHOLOGY Richard J. Crisp
SOCIAL WORK Sally Holland and Jonathan Scourfield
SOCIALISM Michael Newman
SOCIOLINGUISTICS John Edwards
SOCIOLOGY Steve Bruce
SOCRATES C. C. W. Taylor
SOFT MATTER Tom McLeish
SOUND Mike Goldsmith
SOUTHEAST ASIA James R. Rush
THE SOVIET UNION Stephen Lovell
THE SPANISH CIVIL WAR Helen Graham
SPANISH LITERATURE Jo Labanyi
SPINOZA Roger Scruton
SPIRITUALITY Philip Sheldrake
SPORT Mike Cronin
STARS Andrew King
STATISTICS David J. Hand
STEM CELLS Jonathan Slack
STOICISM Brad Inwood
STRUCTURAL ENGINEERING David Blockley
STUART BRITAIN John Morrill
THE SUN Philip Judge
SUPERCONDUCTIVITY Stephen Blundell
SUPERSTITION Stuart Vyse
SYMMETRY Ian Stewart
SYNAESTHESIA Julia Simner
SYNTHETIC BIOLOGY Jamie A. Davies
SYSTEMS BIOLOGY Eberhard O. Voit
TAXATION Stephen Smith
TEETH Peter S. Ungar
TELESCOPES Geoff Cottrell
TERRORISM Charles Townshend
THEATRE Marvin Carlson
THEOLOGY David F. Ford
THINKING AND REASONING Jonathan St B. T. Evans
THOUGHT Tim Bayne
TIBETAN BUDDHISM Matthew T. Kapstein
TIDES David George Bowers and Emyr Martyn Roberts
TOCQUEVILLE Harvey C. Mansfield
LEO TOLSTOY Liza Knapp
TOPOLOGY Richard Earl
TRAGEDY Adrian Poole
TRANSLATION Matthew Reynolds
THE TREATY OF VERSAILLES Michael S. Neiberg
TRIGONOMETRY Glen Van Brummelen
THE TROJAN WAR Eric H. Cline
TRUST Katherine Hawley
THE TUDORS John Guy
TWENTIETH-CENTURY BRITAIN Kenneth O. Morgan
TYPOGRAPHY Paul Luna
THE UNITED NATIONS Jussi M. Hanhimäki
UNIVERSITIES AND COLLEGES David Palfreyman and Paul Temple
THE U.S. CIVIL WAR Louis P. Masur
THE U.S. CONGRESS Donald A. Ritchie
THE U.S. CONSTITUTION David J. Bodenhamer
THE U.S. SUPREME COURT Linda Greenhouse

UTILITARIANISM Katarzyna de Lazari-Radek and Peter Singer
UTOPIANISM Lyman Tower Sargent
VETERINARY SCIENCE James Yeates
THE VIKINGS Julian D. Richards
THE VIRTUES Craig A. Boyd and Kevin Timpe
VIRUSES Dorothy H. Crawford
VOLCANOES Michael J. Branney and Jan Zalasiewicz
VOLTAIRE Nicholas Cronk
WAR AND RELIGION Jolyon Mitchell and Joshua Rey
WAR AND TECHNOLOGY Alex Roland
WATER John Finney
WAVES Mike Goldsmith
WEATHER Storm Dunlop
THE WELFARE STATE David Garland
WITCHCRAFT Malcolm Gaskill
WITTGENSTEIN A. C. Grayling
WORK Stephen Fineman
WORLD MUSIC Philip Bohlman
THE WORLD TRADE ORGANIZATION Amrita Narlikar
WORLD WAR II Gerhard L. Weinberg
WRITING AND SCRIPT Andrew Robinson
ZIONISM Michael Stanislawski
ÉMILE ZOLA Brian Nelson

Available soon:

PHILOSOPHY OF MIND Barbara Gail Montero
INSECTS Simon Leather
PAKISTAN Pippa Virdee
TIME Jenann Ismael
JAMES JOYCE Colin MacCabe

For more information visit our website

www.oup.com/vsi/

David A. Rothery

PLANETS

A Very Short Introduction

OXFORD
UNIVERSITY PRESS

OXFORD
UNIVERSITY PRESS

Great Clarendon Street, Oxford OX2 6DP

Oxford University Press is a department of the University of Oxford.
It furthers the University's objective of excellence in research, scholarship,
and education by publishing worldwide in

Oxford New York

Auckland Cape Town Dar es Salaam Hong Kong Karachi
Kuala Lumpur Madrid Melbourne Mexico City Nairobi
New Delhi Shanghai Taipei Toronto

With offices in

Argentina Austria Brazil Chile Czech Republic France Greece
Guatemala Hungary Italy Japan Poland Portugal Singapore
South Korea Switzerland Thailand Turkey Ukraine Vietnam

Oxford is a registered trade mark of Oxford University Press
in the UK and in certain other countries

Published in the United States
by Oxford University Press Inc., New York

First published 2010
Revised impression published 2021

British Library Cataloguing in Publication Data
Data available

Library of Congress Cataloging in Publication Data
Data available

Typeset by SPI Publisher Services, Pondicherry, India

Printed and bound by
CPI Group (UK) Ltd, Croydon, CR0 4YY

ISBN: 978–0–19–957350–9

Contents

List of illustrations

List of tables

Introduction

In a sonnet of 1816 reflecting upon his experience of reading a new translation of the works of Homer, the young English poet John Keats wrote of 'breathing the pure serene' in 'realms of gold', and continued:

> Then felt I like some watcher of the skies
> When a new planet swims into his ken;
> Or like stout Cortez when with eagle eyes
> He star'd at the Pacific – and all his men
> Look'd at each other with a wild surmise –
> Silent, upon a peak in Darien.

Keats's new planet metaphor was either inspired by Sir William Herschel's sighting of Uranus in 1781, or by discoveries of the first four asteroids (1801–7). Being more recent, the latter would have been fresher in people's memories. A layman such as Keats would have thought of them as new 'planets', although today they are regarded as too small to qualify.

I still travel in 'realms of gold' when I see Saturn with my own eyes through even a small telescope, though the novelty has somewhat faded when it comes to seeing a newly discovered remote ice-ball as a single pixel in a digital image, or a hint of a Jupiter-sized companion to another star expressed as a minute wobble in the star's position.

However, for me, the true 'Cortez experience' recurs whenever I see a new planetary landscape (in some cases, a cloudscape) unfold before me on images sent back from a visiting spacecraft. Exploration of our Solar System has reached a stage that allows us to appreciate other planets and their large satellites as worlds, endowed with geographies, geologies, and meteorologies as complex and fascinating as those of our own planet, Earth. Many of them are places that you and I could, in principle, visit. They are not generally suitable for a picnic, but we could at least jump up and down, scoop up handfuls of dirt, climb a hill, or slither down into a valley. Some are even places where life might be found.

In this book, I will share with you what is known of the origin, evolution, and especially the present-day condition of the planets in our Solar System. Here, astronomers now officially recognize only eight planets (Pluto has been demoted, as I will describe), but there are plenty of other bodies big enough to behave like planets so far as geologists like myself are concerned. These are fascinating, so I will not ignore them, though they are too numerous to treat individually.

Finally, I will turn to 'exoplanets', which are planets orbiting other stars. The first was discovered as recently as 1995, and by now thousands have been documented. We cannot see them in any detail, but we do have enough information to make some comparisons between the layouts of those exoplanetary systems and the Sun's family.

Chapter 1
The Solar System

Planets in history

Before the curses of light pollution and smog, people were more familiar with the night sky than they tend to be today. Planets in the sky were recognized as special by ancient cultures, because they are 'wandering stars' that migrate against the background of the 'fixed' stars. Five planets have been known since antiquity: Mercury, Venus, Mars, Jupiter, and Saturn – which are the only ones bright enough to come to the attention of the unaided eye. Of course, the Sun and Moon were obvious too, but the 'planets' appear as wandering points of light, whereas the Sun and Moon show disks and tended to be regarded differently. Throughout most of humankind's existence, the Earth was imagined to be the centre of creation, unrelated to objects in the sky, so it was not thought of as a planet.

The intellectual leaps that recognized that the Earth is a ball of rock going round the Sun, that the planets do likewise, and that the Earth is just one of their number were a long time coming. The process was slow, and there were many false dawns. During the 5th century BC, the ancient Greek philosopher Anaxagoras correctly surmised that the Moon is a spherical body reflecting the light of the Sun, and he was sent into exile on account of his beliefs. In the succeeding centuries, various Chinese astronomers developed similar ideas, but the idea of the Moon as a globe probably did not embed itself into popular consciousness until its appearance through a telescope became known during the 17th century.

As for the planets, they were generally regarded as points of light going round the Earth, until the counterintuitive 'heliocentric' view with the Sun as the centre of motion became accepted. The earliest written suggestions that the Earth goes round the Sun occur in Indian texts dating from the 9th century BC, but despite this and subsequent independent suggestions, notably by Hellenic and Islamic sages and eventually by Nikolas Copernicus in 1543, the concept did not achieve ascendancy until the 18th century. Partly on account of his advocacy of the heliocentric theory, Galileo Galilei (who through his telescope had seen mountains on the Moon, the phases of Venus, and four tiny moons orbiting Jupiter) was held under house arrest from 1633 until his death in 1642.

Simply by revealing the planets as tiny but discernible discs, whereas the stars remained as points of light, use of the telescope from the start of the 17th century onwards marked planets as fundamentally different to stars, and eased the path to regarding them as worlds comparable to our own. Incidentally, we now know that stars are much bigger than planets, but (except for the Sun) they are so very much more distant that only in a few cases can even the most sophisticated of modern telescopes show any surface details (on photographs, bright stars look bigger than faint stars, but that is just an optical effect – the brightness is being smeared out).

Kepler's laws of planetary motion

The planets slotted into their rightful place in human comprehension thanks to Johannes Kepler's (1609) realization that the planets (including the Earth) travel round the Sun in paths (orbits) that are ellipses rather than perfect circles, coupled with Isaac Newton's (1687) insight into gravity that explained this motion. Then their distances and sizes relative to the Earth could begin to be deduced.

An ellipse is what you might think of as an 'oval'. Mathematically, it is defined as a closed curve drawn about two points (the foci of the ellipse) such that the sum of the distances from each focus to any point on the curve is identical. A circle is a special kind of ellipse in which the two foci coincide, at the circle's centre. The further apart the foci, the more elongated, or 'eccentric', the ellipse. Kepler deduced that planets follow elliptical orbits, with the Sun at one focus of each ellipse (the other focus being empty). The point on an orbit closest to the Sun is called 'perihelion' (Greek for 'closest to the Sun'), and the point furthest away is called 'aphelion' (Greek for 'furthest from the Sun'). Planets' orbits are not strongly eccentric, and if you see them drawn in plan view they look very much like circles. For example, when Mars is at aphelion its distance from the Sun is less than 21% greater than when it is at perihelion, and for the Earth the difference is only 4%.

Kepler is justly famous for his three laws of planetary motion. Kepler's First Law is simply the statement that each planet moves in an elliptical orbit, with the Sun at one focus. The Second Law describes how the speed of a planet varies around its orbit: a planet moves faster the closer it is to the Sun (for reasons subsequently explained by Newton's theory of gravity) such that an imaginary line linking the planet to the Sun sweeps out an equal area in equal time. Kepler's Third Law relates a planet's orbital period (how long it takes to complete a circuit round the Sun) to its average distance from the Sun: the square of the orbital period is proportional to the cube of the average distance. The average distance from planet to Sun turns out to be equal to half the length of the orbital ellipse's long axis (its 'semi-major axis') or, if you prefer, half the straight-line distance between perihelion and aphelion.

Kepler's laws of planetary motion enabled precise calculation of the sizes of the orbits of other planets, with an accuracy limited almost entirely by the uncertainty in how well the size of the Earth's orbit could be measured. Even as long ago as 1672, simultaneous observations of Mars from widespread locations enabled the Earth–Sun distance to be measured as about 140 million kilometres, remarkably close to the correct value of 149,597,871 kilometres. Observations of the transit of Venus across the Sun's disc in 1761 and 1769 (the latter requiring Captain Cook to station himself in Tahiti) produced a revised estimate of 153±1 million kilometres. Despite these and other scientific advances, which continued to strengthen a fully self-consistent and elegant model of the Solar System's scale and nature, a papal ban on printing 'heliocentric' books in Rome remained unrevoked until 1822.

You would be excused for thinking that once the distance to a planet has been established, working out its size would be trivial. However, the smallness of a planetary disc through even a large telescope, coupled with the shimmering of the Earth's atmosphere, leads to significant uncertainty in measuring the angular size of the planet (in other words, how big it appears). For example, when he discovered Uranus in 1781, William Herschel's measurement of its disc was 8% too large. Rather than trying to measure how big a planet looks, the most precise telescopic way to determine its size is to time how long it takes to pass in front of a star. Such 'occultations' are rare events, but by the close of the 19th century the sizes of the planets had been determined with considerable accuracy (Table 1).

Herschel discovered Uranus by accident, but Neptune was located in 1846 as a result of a deliberate search, guided by slight perturbations in the orbit of Uranus (distorting it from a perfect ellipse) that could be best explained by the gravitational influence of an unseen outer planet. When it had been documented for long enough, the orbit of Neptune in turn seemed to show perturbations pointing to a further undiscovered planet. This triggered a search that found Pluto in 1930. At first, astronomers

Table 1 The sizes of the planets (equatorial diameters)

Planet	Value published in 1894*	Correct value
Mercury	4,720 km	4,879 km
Venus	12,600 km	12,104 km
Earth	12,756 km	12,752 km
Mars	6,760 km	6,794 km
Jupiter	142,000 km	142,980 km
Saturn	119,000 km	120,540 km
Uranus	53,600 km	51,120 km
Neptune	48,500 km	49,540 km

* C. Flammarion, *Popular Astronomy* (Chatto and Windus, Piccadilly)

assumed that this newly hailed ninth planet must be similar in size and mass to Uranus and Neptune. However, by 1955 it had been shown that Pluto could be no larger than the Earth; in 1971 the estimate was revised downwards to the size of Mars; and in 1978 its surface was found to be dominated by highly reflective frozen methane which meant that its physical size had to be even smaller to remain consistent with its total brightness. We now know that Pluto's diameter is only 2,377 kilometres, so its size is smaller (and, in fact, its mass is *much* smaller) even than Mercury. The apparent perturbations in Neptune's orbit that, rather fortunately, inspired the search for Pluto are now attributed to observational inaccuracies.

Pluto lost its status as an officially recognized planet in 2006. That was a contentious move, though in my opinion the right one. Before describing how this came about, I will review the nature of the Solar System as it is now understood.

A review of the Solar System

The Sun

In the centre of the Solar System is the Sun, which is a fairly ordinary star, powered by the conversion of hydrogen into helium by nuclear fusion in its core. The Sun's diameter is 109 times and its mass is nearly 333,000 times greater than the Earth's. It contains about 740 times more mass than everything else in the Solar System put together. Consequently, the Sun's gravity is so dominant that objects in the Solar System orbit the Sun in almost the perfect ellipses recognized by Kepler. Perturbations to a planet's orbit caused by other planets are tiny, though they can be measured.

The planets

Table 2 summarizes some basic properties of the planets, quoted relative to the Earth to avoid very large numbers. Distance from the Sun is quoted in 'Astronomical Units', abbreviated as AU, defined as the average Earth–Sun distance. This is fairly simple to remember as (near enough) 150 million kilometres. A planet's orbital period is how long it takes to complete one circuit round the Sun, which is of course its own 'year'. The orbital periods and distances from the Sun in this table are related to each other by Kepler's Third Law. Conveniently, this means that the square of any planet's orbital period (in Earth-years) is equal to the cube of its average distance from the Sun (in AU). The Earth's mass is very nearly 6 million billion billion kilograms (or 6 thousand billion billion tonnes), hence the convenience of comparing other planets to the Earth rather than quoting standard scientific units such as kilograms, seconds, and metres.

Rotation period is how long it takes a planet to spin once on its axis. For a rapidly spinning planet, this is almost the same as the time from one sunrise to the next (the planet's own 'day length'), but the relationship is not exact because a planet's orbital motion continuously changes the direction between planet and Sun. The Earth's rotation period is 23 hours and 56 minutes, but it takes exactly 24 hours to rotate far enough to bring the Sun back to the same point in the sky. From a planet's perspective, the Sun migrates completely round the sky during the course of a single orbit, in addition to the changing direction towards the Sun from

Table 2 Some properties of the planets compared. Distance from the Sun refers to average distance. Years and days are Earth-years and Earth-days. See Table 1 for sizes

Planet	Distance from Sun / AU	Orbital period / years	Mass (relative to Earth)	Rotation period / days
Mercury	0.387	0.241	0.055	58.6
Venus	0.723	0.615	0.81	243.0
Earth	1.0	1.0	1.0	1.0
Mars	1.52	1.88	0.11	1.026
Jupiter	5.20	11.86	318	0.410
Saturn	9.58	29.46	95.2	0.444
Uranus	19.1	84.01	14.5	0.718
Neptune	30.0	164.8	17.2	0.768

any point of the planet's surface caused by the planet's rotation. A planet whose rotation had become tidally locked so that it rotated exactly once per orbit (synchronous rotation) would keep one face permanently towards the Sun. Mercury does not quite do this, but rotates exactly *three* times during the course of *two* orbits, as a result of which it turns *relative to the Sun* once per two orbits, so its day is twice as long as its year.

There is a change in character between the four inner planets and the four outer ones. The inner planets (Mercury, Venus, Earth, and Mars) are relatively small and low in mass compared to the outer four (Jupiter, Saturn, Uranus, and Neptune). There is also a contrast in their densities, the inner planets being denser than the outer ones. The inner planets are called the 'terrestrial planets', signifying that they are all 'Earth-like'. The outer four are the 'giant planets'. Some call them 'gas giants' to reflect the fact that they have so much hydrogen and helium. Others reserve that particular term for just Jupiter and Saturn, which are more gassy than the other two, though even those each contain more than one Earth-mass of gas.

Figure 1 is a map of the Solar System, showing orbits to scale, except that the orbits of Venus and Mercury are too small to

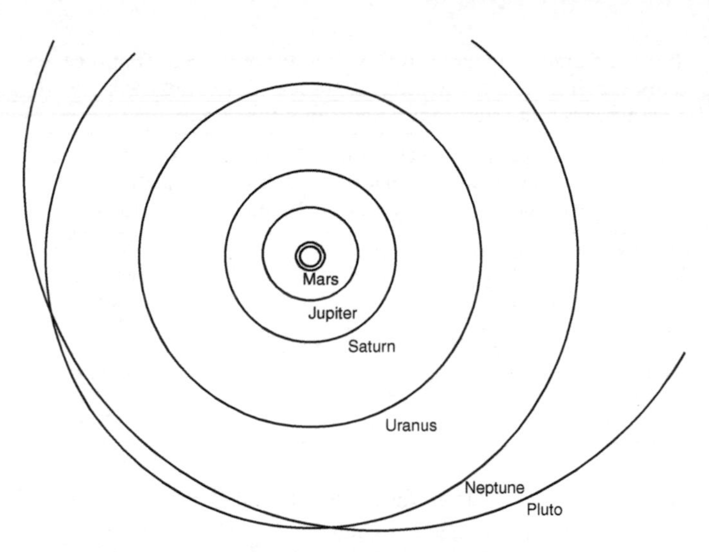

1. Map of the Solar System, showing planetary orbits at the correct relative sizes. Orbits are only slightly eccentric, so look virtually indistinguishable from circles. The unlabelled circle inside Mars's orbit is the Earth's orbit, not the Sun! The orbits of Venus and Mercury are too small to include. Pluto is not a planet, but its orbit is shown because it is representative of a large number of small bodies beyond Neptune's orbit

include. Part of Pluto's orbit is shown, to help with discussion later. Something that I have not yet mentioned, but without which such a map could not be drawn, is that planetary orbits all lie approximately in the same plane. Relative to the Earth's orbit, which makes a convenient reference plane known as the 'ecliptic', Pluto's orbit is inclined at 17.1°, Mercury's at 7.0°, Venus's at 3.4°, and all the others at less than 3°.

When Pluto is near perihelion, it is inside the orbit of Neptune, but there is no prospect of them colliding. Their differing orbital inclinations prevent their paths from intersecting, and moreover Neptune is always on the opposite side of the Sun whenever Pluto passes inside Neptune's orbit. This is possible because for every three orbits completed by Neptune, Pluto completes exactly two. Such a relationship is referred to as 3:2 orbital resonance.

As well as their orbits being nearly coplanar, every planet goes the same way round the Sun: they travel anticlockwise as seen from an imaginary vantage point far above the Earth's north pole. Anticlockwise motion is also manifested in the direction in which each planet except Venus and Uranus rotates on its axis. Because anticlockwise motion is so common, it is called 'prograde'. Clockwise orbital motion or rotation is regarded as backwards and is referred to as 'retrograde'.

With the exception of Uranus, the axis about which each planet rotates is less than 30° away from being at right angles to its orbital plane. Mercury is nearly 'perfect', with a tilt of only 0.1°, whereas Earth's axis is tilted at 23.5°. The direction in which a planet's axis points and the amount of tilt both vary when measured over tens of thousands of years, but they are effectively constant on the timescale of a single orbit. Axial tilt is why planets have seasons; on Earth, summer occurs in the northern hemisphere during that part of the orbit when the north end of the Earth's axis is tilted *towards* the Sun, and northern winter is six months later when the Earth is on the other side of the Sun, so that the north end of the axis is tilted *away* from the Sun. Of the two planets that don't conform, Venus's axis is tilted at only 2.7° but it rotates very slowly in the retrograde direction (giving it a day length of 116.7 Earth-days), whereas Uranus's axis is tilted by 82.1° with rapid retrograde rotation. Uranus probably suffered a catastrophe that knocked it over, having started with prograde rotation that became tipped over by 97.9° (97.9° being 180° minus 82.1°). This would result in the present situation without calling on a separate event to reverse its direction of spin.

Satellites of planets

All planets except Mercury and Venus have satellites, or 'moons'. These are smaller bodies close enough to orbit the planet rather than the Sun. Strictly speaking, a planet and its satellite each orbit their common centre of mass (or 'barycentre'). However, planets are so much more massive than their satellites that their barycentre is inside the planet, and it is usually perfectly adequate to regard satellites as going round their planet. Most satellites' orbits lie close to their planet's equatorial plane and almost all the large ones have prograde orbits, which is defined as orbiting in the same direction as the planet's spin.

The Earth's satellite is, of course, the Moon (with a capital M). This is exceptional in being relatively large in comparison to its planet, having a diameter 27% and a mass 1.2% of the Earth's. By coincidence, the Moon's size and distance from Earth are such that it appears almost the same as the Sun, which is much larger but correspondingly further away. When the Moon passes exactly between the Earth and the Sun, it hides the Sun's disk, causing a solar eclipse. If the Moon's orbit round the Earth were exactly coplanar with the Earth's orbit, there would be an eclipse every lunar orbit (that is, every month). However, the Moon's orbit is inclined at 5.2° to the ecliptic, so eclipses are rare, occurring only when the Moon happens to pass between the Earth and the Sun at one of the two points where its orbit crosses the ecliptic. Unravelling the cyclic nature of these events and predicting when eclipses would occur (though without fully understanding the reasons) was one of the great achievements of Babylonian astronomers about 2,600 years ago. Lunar eclipses, when the Moon passes into the Earth's shadow, are controlled by the same cycle, but are more common because the Earth's shadow is considerably bigger than the Moon.

Mars has two tiny satellites. Jupiter has four exceeding 3,000 kilometres in diameter (which are those that Galileo discovered), plus at the last count 74 others less than 200 kilometres (most less than 4 kilometres) across. Saturn has slightly fewer (62) known satellites, though only one of them rivals Jupiter's largest. Uranus has five satellites between 400 and 1,600 kilometres across and 22 known smaller ones, and Neptune has one large satellite and thirteen known small ones. Most of the small (few kilometres across) outer satellites of Jupiter and Saturn were discovered using telescopes (rather than by visiting spacecraft), and more tiny satellites of the giant planets surely remain to be found, especially at Uranus and Neptune, where detection by telescope is especially hard for two reasons: they are further from the Sun and so are less brightly lit, and they are further from the Earth and so would look fainter even if they were equally well lit.

The larger satellites are geologically very interesting and I say more about them later, but all satellites are useful for the planetary scientist because they enable their planet to be weighed. The orbital period of a satellite depends only on its average distance from the planet's centre and their combined mass (which can be calculated using Newton's gravitational elaboration of Kepler's Third Law). Because satellites are so much smaller, the mass of the planet dominates almost entirely, in the same way that planets' orbits round the Sun depend on distance and solar mass.

Asteroids, trans-Neptunian objects, and comets

This book is about planets, rather than the whole Solar System, but it is worth noting that objects of other types vastly outnumber the planets and their satellites combined, although admittedly these are small and their total mass is relatively insignificant. Although planetary scientists have come to realize that the boundaries are somewhat blurred, these 'junk' objects can be divided into three broad classes: asteroids, trans-Neptunian objects, and comets.

Asteroids range downwards in size from 950 kilometres across (the diameter of Ceres, the largest example), with no lower limit. Asteroids only a few tens of metres across have been detected as they pass close by the Earth, and the remains of smaller ones that fall to the ground can be found as meteorites. Formerly assumed to be fragments of a destroyed planet, we now think that asteroids never belonged to a planet-sized object. The total mass of all the asteroids is probably less than a thousandth of the Earth's mass. Some have clearly experienced mutual collisions, as attested by their irregular shapes.

Without exception, asteroid orbital motion is prograde. Most have orbital inclinations of less than 20°, but eccentricity is typically greater than for planets. The orbits of most asteroids lie between those of Mars and Jupiter (the so-called 'asteroid belt'), but some come much closer to the Sun, passing inwards of the Earth and even (in a handful of examples) inwards of Mercury. A few asteroids are known orbiting beyond Saturn. Like the meteorites derived from them, most asteroids are stony or carbonaceous in composition, but some are made of iron and nickel. So far as we can tell, asteroid composition tends to be less stony and more carbonaceous and eventually icier with distance from the Sun.

Beyond the orbit of Neptune, between about 30 and 55 AU from the Sun, small icy bodies become common, and there are several that exceed the largest asteroids in size. This region is usually called the 'Kuiper belt', named after the Dutch-American Gerard Kuiper who predicted it in 1951 as a zone where icy lumps should be left over from the birth of the Solar System. An Irishman, Kenneth Edgeworth, said much the same in a more obscure journal in 1943, so some prefer to call this the 'Edgeworth-Kuiper belt'. The first Kuiper belt object to be discovered and recognized as such was found in 1992, but now many hundreds of them have been catalogued, and it has become clear that Pluto should be numbered among them. Similar objects with perihelion not far beyond Neptune's orbit but reaching about 100 AU at aphelion are counted as 'Scattered Disk' objects. Together with the Kuiper belt, these make up a family called 'trans-Neptunian objects', or TNOs, all in prograde orbits. The total mass of TNOs is probably around 200 times that of the asteroid belt (one-fifth of an Earth-mass), and in total there may be nearly 100,000 bodies more than 100 kilometres in size. One 'Scattered Disk' object discovered in 2005 and subsequently named Eris is only marginally smaller than Pluto. We can be more confident about their masses because they both have satellites with well-documented orbits, showing that the mass of Eris is 28% greater than that of Pluto.

Comets have been known since antiquity because a comet can briefly look very spectacular, thanks to the development of a tail of gas and dust that can stretch across the sky when the comet passes close to the Sun. However, the solid part of a comet is just a chunk of dusty ice (famously described as a 'dirty snowball'), only a few kilometres across in most cases. A comet spends most of the time far from the Sun, and develops a tail only when it passes close enough for the Sun to warm it. This happens rarely because comets have extremely eccentric orbits with perihelion usually inside the Earth's orbit but aphelion near or well beyond Jupiter's orbit. Some come from so far beyond that their orbits look like parabolas (infinitely long ellipses), and make only one passage close to the Sun throughout recorded history. Those are 'long-period' comets, and appear to have been dislodged from an ill-defined shell surrounding the Sun at about 50,000 AU known as the Oort Cloud. In contrast, 'short-period' comets probably originated as Scattered Disk objects that were perturbed into an eccentric orbit with a small perihelion distance by a close encounter with a fellow object. Those with orbital periods of hundreds of years still have their aphelion in the Scattered Disk, but aphelion can be nudged closer to the Sun as a result of a close encounter with a giant planet. For example, Halley's comet has aphelion near Neptune's orbit, and an orbital period of 75 years, whereas Encke's comet has aphelion near Jupiter's orbit and a period of only 3.3 years. Comets lose mass by degassing every time the Sun's heat warms them, so after fewer than a thousand perihelion passages a comet is probably reduced to an inert aggregate of ice-free rock and dust, hard to distinguish from an asteroid.

As you might expect, given their source, the orbits of 'short-period' comets are prograde and tend to be close to the ecliptic. However, no such restriction applies to long-period comets whose orbits can be highly inclined or even retrograde.

What is a planet? How Pluto came to be thrown out of the club

The first TNO to be discovered was Pluto, in 1930. Even after Pluto's small size became apparent (and subsequently, thanks to the discovery of its largest satellite in 1978, its small mass), people tended to think of Pluto as the ninth planet. However, when the known population of the Kuiper belt blossomed into hundreds of objects, several of which rivalled Pluto in size, it became increasingly anomalous to classify Pluto as a planet and yet other Kuiper belt objects as something different. When Eris was confirmed to be more massive and similar in size to Pluto, then logically either all the large TNOs had to be called planets, or none of them. However, many people argued to retain Pluto as a planet on sentimental or historical grounds.

Decision-making was hampered by the fact that the term 'planet' had never been fully defined. Eventually, at a meeting of the International Astronomical Union held in Prague in 2006, during which passions ran high, delegates voted to accept some definitions that have largely settled the issue. Two criteria for planethood were non-controversial: the IAU ruled that, firstly, a planet must have sufficient mass for its self-gravity to overcome 'rigid body forces' so that it assumes a hydrostatic equilibrium (nearly round) shape, and secondly that it must be in orbit about the Sun. This second criterion rules out large satellites such as our own Moon.

A third criterion was the crucial one. It states that to be counted as a planet, a body must have 'cleared the neighbourhood around its orbit' of everything except much smaller objects. This is the test that Pluto fails. Pluto has not cleared its neighbourhood, because it shares it with many bodies of similar size and indeed also with the vastly more massive Neptune. On the other hand, Neptune *does* pass the test, because it is many thousands of times more massive than anything else in the same orbital region (such as Pluto).

Having taken the bold but entirely logical step to expel Pluto from the planetary club, the IAU seems immediately to have regretted it, and invented not one but two new classes for it to belong to. At the 2006 Prague meeting, the newly coined term 'dwarf planet' was defined as 'a celestial body that is in orbit round the Sun, has sufficient mass for its own gravity to pull it into a nearly round shape, has *not* cleared the neighbourhood of its orbit, and is not a satellite'. Determining whether or not shape is 'nearly round' is difficult to do remotely, and controversial to define, but in adopting this definition the IAU gave Pluto, Eris, and Ceres (the largest asteroid) the consolation prize of being called 'dwarf planets'. At the time, it was acknowledged that other large TNOs would be ranked as dwarf planets when they had been adequately measured, and sure enough in 2008 a Kuiper belt object named Makemake (pronounced as four syllables), estimated to be about two-thirds the size of Pluto, was judged to pass the shape test and was admitted as a fourth dwarf planet, closely followed by a fifth called Haumea.

The IAU seemingly came to regret having lumped Pluto-like objects in the same category as Ceres, and in 2008 invented a new term, 'Plutoid', to denote trans-Neptunian dwarf planets. Ceres is thus the only dwarf planet that is not a Plutoid, and there is surely no undiscovered asteroid large enough ever to join it in that category. However, there are probably numerous undiscovered or inadequately documented large TNOs that will join Pluto, Eris, Makemake, and Haumea as both Plutoids *and* dwarf planets. Incidentally, Eris is named (appropriately, considering the controversy that it caused) after a classical Greek goddess of discord, whereas Makemake and Haumea are named after Pacific island fertility deities.

How it all happened

Growing planets

Until recently, it would have been possible to argue that planets are rare in the cosmos, but it now seems clear that planets are a usual by-product of star formation. The existence of our Solar System is thus a consequence of the origin of the Sun itself.

A star is believed to form when a vast interstellar cloud made mostly of hydrogen, but mixed with a few other gases and tiny solid particles referred to as dust, contracts under the influence of its own gravity. As the cloud contracts, most of the matter becomes concentrated into the centre, in a body that becomes increasingly hot because of the gravitational energy converted to heat by the process of infall. Eventually, the central pressure and temperature rise so high that hydrogen nuclei begin to fuse together to make helium, at which stage the central body can be called a star. The planets grow from some of the material left behind during the final stages. Conservation of angular momentum causes any slight initial rotation of the cloud to speed up during contraction, and matter not incorporated into the star becomes concentrated into a disk in the star's equatorial plane, rotating in the same direction as the star.

This rotating disk is where planets form. The one that gave birth to our Solar System is referred to as the solar nebula, 'nebula' being Latin for 'cloud' and used by astronomers to denote any large mass of gas and/or dust in space. There are strong reasons for believing that the solar nebula's composition was about 71% hydrogen, 27% helium, 1% oxygen, 0.3% carbon, and 0.1% each of nitrogen, neon, magnesium, silicon, and iron. Almost all the original dust in the solar nebula was probably vaporized by heat from the young Sun, but soon conditions in the nebula became cool enough for new dust grains to condense, not as individual elements but as compounds produced by chemical combination.

Helium does not combine into chemical compounds, so the most abundant compounds that could condense involve either hydrogen or oxygen.

Thanks to the available elements and the local temperature and pressure, oxygen was able to bond with silicon and various metals to form a range of compounds called silicates in the inner part of the nebula. These are common minerals on Earth that crystallize when molten rock cools, but in the solar nebula they grew directly from gas. Hydrogen was incorporated into solid particles only where the temperature was low enough for hydrogen-bearing compounds to form, and for most purposes this seems to have been beyond about 5 AU from the Sun. At and beyond this so-called 'ice line', water (made of hydrogen plus oxygen) could condense as specks of ice. Further from the Sun, more volatile compounds formed where hydrogen bonded with carbon to make methane and with nitrogen to make ammonia, and carbon with oxygen to form either carbon monoxide or carbon dioxide. At about 30 AU, it was cold enough for nitrogen to condense as solid particles of pure nitrogen. By one of the quirks of planetary science vocabulary, *any* solid made of water, methane, ammonia, carbon monoxide, carbon dioxide, or nitrogen (or indeed any mixture of these) is referred to as 'ice', recognizing similarities in origin and properties. This means that, to avoid ambiguity, planetary scientists have to specify 'water-ice' when referring specifically to frozen water – a complication that rarely arises on Earth, where temperatures are too high for compounds more volatile than water to freeze naturally.

Condensation happened in such a way that the first dust grains – microscopic specks made of silicates close to the Sun and ices (plus some leftover silicates) further from the Sun – did not grow as dense, rigid specks. Instead, they had intricate 'fluffy' shapes, and when these bumped into each other, they tended to stick together rather than bouncing apart. Within as little as ten thousand years after the onset of condensation, the particles could have grown into globules about a centimetre across through the combined effects of continuing condensation and accretion (sticking together) when they collided. After perhaps 100,000 years, the Solar System would have consisted of hordes of bodies about 10 kilometres across, dubbed 'planetesimals'. These were all swirling round the Sun in the same, prograde, direction and enclosed in a diffuse haze made of the remaining gas and dust.

We know how long ago this happened, because some of the earliest grains survive unaltered inside meteorites. We can measure radioactive decay products within them to work out their age, which is a particularly memorable number: 4.567 billion years. The most 'primitive' meteorites, which are fragments of small planetesimals that never suffered heating or alteration, are called 'carbonaceous chondrites' and are our most direct evidence of conditions in the early Solar System.

Hitherto, collisions had been essentially a matter of chance, but once planetesimals reached about 10 kilometres in size, the greater gravitational pull of the largest ones was able to make itself felt. These suffered more frequent collisions, so their rate of growth outpaced that of the others. Within a few more tens of thousands of years, the largest planetesimals had grown to a thousand or so kilometres across, gobbling up most of the smaller ones in the process.

These large planetesimals are dignified by a new name: 'planetary embryos'. Maybe a few hundred were formed in the inner Solar System. They would have been massive enough for their own gravity to pull them into spherical shapes. They may have been hot enough internally for melting to occur, allowing iron to sink inwards to form a distinct core, but that is largely immaterial because of what happened next.

These planetary embryos were what the terrestrial planets formed from. Now that the majority of the small stuff was gone, significant growth could happen only when two embryos smashed together. Such a collision is referred to as a 'giant impact', and liberates enough heat to largely melt the merged body formed by the collision. Imagine a sphere of molten rock, glowing red hot except for a few rafts of chilled clinker on its surface, with deep inside a 'rain' of iron droplets settling inwards through the silicate magma to accrete onto the central core, and you should have a picture in your mind that conveys the state of a planetary embryo in the aftermath of a giant impact.

This assumes that the impact doesn't smash both bodies to smithereens, but inevitably a certain amount of debris will be thrown out to space as ejecta from the collision. It probably took about 50 million years to build up an Earth-sized planet by serial giant impacts between planetary embryos. Because of the randomness of the collisions and the complex 'family tree' of giant impact collisions between bodies that themselves had been formed by giant impacts, it is meaningless to regard any single embryo early in the process as 'the proto-Earth' or 'the proto-Venus'.

Beyond the orbit of Mars, the gravitational effect of the young Jupiter was strong enough to stir the rocky planetesimals into more eccentric orbits, so that mutual collisions were often too violent to allow growth by accretion. Instead, fragmentation was a common outcome, so large planetary embryos that might have eventually collided to produce a fifth terrestrial planet were unable to grow here. Today, in that region, we find most of the asteroids, representing only a tiny fraction of the mass that once existed there. Jupiter scattered the majority into markedly eccentric orbits, so that eventually most collided with Jupiter or another giant planet, or were ejected from the Solar System entirely.

The bodies from which the giant planets formed had a high proportion of ice as well as rock within them. There, beyond the 'ice line', the growing planets had much more material to draw upon. The role of embryo–embryo collisions is uncertain, and so is the mechanism by which they acquired so much gas. One theory is that after they had exceeded 10 or 15 Earth-masses, their gravitational pull was sufficient to scavenge huge quantities of whatever gas survived in the remaining nebula, and so their rocky and icy kernels became encased within deep gassy envelopes. Another school of thought holds that gravitational instabilities in the nebula caused each giant planet to grow inside a particularly dense knot, where the gas was naturally confined about the growing planet.

Opinion is also divided over the relative rates of planetary growth in the inner and outer parts of the Solar System, and it is unclear whether Jupiter formed before or after the Earth and Venus. However, if they grew by embryo–embryo collisions, Saturn, Uranus, and Neptune must have grown more slowly than Jupiter because collisions should be less frequent with increasing distance from the Sun.

Scavenging of gas from the nebula was terminated when the Sun entered its 'T Tauri' phase, named after the star T Tauri which is undergoing this process today. For perhaps 10 million years, a strong outflow of gas from the star, called the 'T Tauri wind', blows away all the remaining gas and dust. A likely reason for Uranus and Neptune having proportionally less gas than the other giant planets is that they took longer to grow, leaving less time to collect gas before the T Tauri wind put an end to the process.

Migrating planets

A further matter of debate concerns ways in which orbits can change over time and the extent to which this happened, particularly among the giant planets. Until the solar nebula was dispersed, gravitational interactions between nebular material and large orbiting bodies would tend gradually to decrease the radius of their orbits, causing embryos and young planets to migrate inwards. After nebular dispersal, gravitational interactions between planets and smaller bodies could have played an even more dramatic role. Some suggest a period of half a billion years or so when the outermost giant planet was deflecting the orbits of outlying icy planetesimals inwards, where they would be likely eventually to be nudged further inwards by interaction with the next giant planet, and so on until they passed close enough to Jupiter for Jupiter to fling them outwards. These out-flung icy planetesimals could be the origin of today's Oort Cloud. Jupiter must have moved fractionally closer to the Sun each time it flung a body outwards, but conversely the other giant planets would have been nudged outwards each time one of them swung a lump of ice inwards. This story has Jupiter migrating inwards, while Saturn, Uranus, and Neptune migrated outwards. It is even possible that Uranus and Neptune swapped places (providing an opportunity for Uranus's axis to become tipped over into its present state). Today's TNOs are those that survived beyond the zone swept clear during Neptune's outward march.

Please do not form the impression that the orbit of a planet is capable of changing either rapidly or dramatically. Claims that Venus and/or Mars passed close to the Earth during biblical times, triggering various myths and natural disasters, are completely untenable. The outer planet migrations I have described happened extremely slowly, and as a result of cumulative interactions with nebular gas and with vast numbers of small bodies that are no longer available.

Nevertheless, the planets and their mutual gravitational pulls are continuously changing configuration. Chaos theory says it is therefore impossible to predict planetary positions more than a few million years ahead. However, it *can* be shown that the Solar System is sufficiently stable that no planet is likely to collide or be ejected in the next few billion years. We are probably safe for at least 5 billion years, which is when astronomers expect the Sun to swell up into a red giant, whereupon the wanderings of Mars will be the least of the problems faced by any far future Earthlings.

Why all the satellites?

By now, you should not be surprised that there is no straightforward answer to the question of whether satellites somehow grew alongside their planets or were acquired later. The large prograde satellites of the giant planets are the easiest to explain. They are thought to have formed within a cloud of gas and dust surrounding each giant planet as it grew, rather like a miniature version of the solar nebula. Tiny prograde satellites only a few kilometres in size orbiting close to giant planets are probably fragments of larger satellites that came too close and broke apart. The outer satellites of giant planets are mostly in retrograde orbits, and these are probably captured bodies that began as asteroids, TNOs, or comet nuclei.

Theoretically, it is almost impossible for a planet to capture a passing visitor into orbit about itself. An incoming smaller body will be swung past a planet by the pull of its gravity, but it can't easily be slowed down enough to be captured into orbit. However, if the visitor is a double object, one of the pair can be captured by transfer of momentum to the other member, which will scoot away even faster after the encounter. A currently favoured explanation for Neptune's large retrograde satellite, Triton, is that Triton was formerly half of a double TNO that strayed close to Neptune. This seems plausible, because several TNOs are known to be twin bodies. Mind you, it leaves unresolved the issue of why so many TNOs (and indeed asteroids too) have satellites in the first place.

The Earth's Moon has a different explanation, and seems to have condensed from debris thrown into orbit about the Earth by the final embryo–embryo collision of the series by which the Earth grew. The two tiny satellites of Mars (Phobos and Deimos) are asteroids, whose capture into close circular orbits is not understood.

Collisions and the cratering timescale

Although collisions between substantial objects are now extremely rare, there is still a large number of small objects that could eventually collide with a planet. Until about 3.9 billion years ago (an epoch called the 'late heavy bombardment'), the rate at which asteroids and comets were hitting planets was much higher than today. Impact craters of that age are well preserved on the Moon (Figure 2), though cratering has continued at a slower rate ever since. An impact crater forms on a solid body when something hits it at a few tens of kilometres per second, and is excavated by shockwaves that radiate from the point of impact. Craters are circular, except for rare examples when the impacting body arrives at a grazing angle.

There is a well-understood hierarchy of crater morphologies depending on diameter, and which can be reproduced experimentally and in computer models. On the Moon, craters from microscopic size up to 15 kilometres across have simple bowl shapes. Then up to about 140 kilometres diameter, craters do not become deeper but have flat floors, and usually a central peak formed by rebound immediately after excavation. There is a nice example near the top of Figure 2. Larger craters may have a group of central peaks, and then craters bigger than 350 kilometres take the form of two or more concentric rings. The transitions from one type to another occur at slightly smaller diameters on bodies with stronger gravity.

Earth's cratering record is poorly preserved, because it is an active planet where processes that erase or bury craters almost keep pace with the rate at which craters are formed. Fortunately, the vast tracts of ancient terrain surviving on the Moon allow us to count the density of impact craters on surfaces whose ages are known, thanks to datable samples returned to Earth by the Apollo programme of manned lunar landings, supplemented by a few Soviet unmanned sample-return missions. By this means, we know the date of the late heavy bombardment and also the

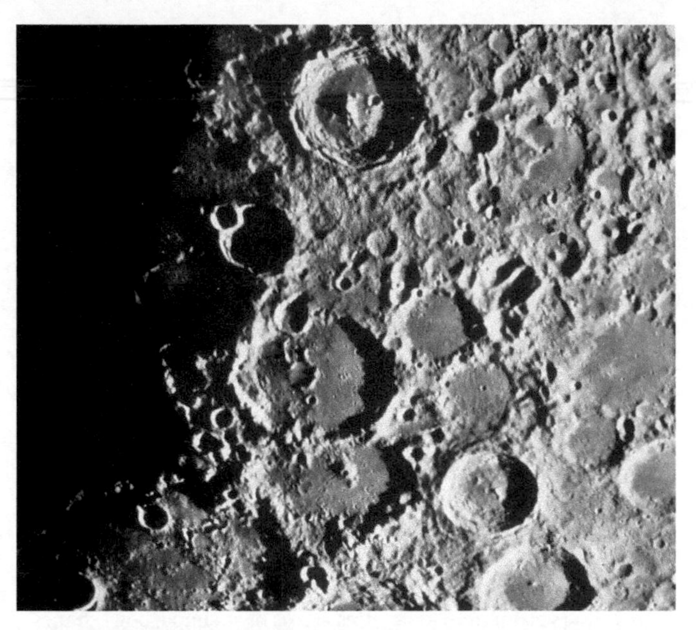

2. A 470-kilometre-wide view of a heavily cratered region of the Moon. Most of these craters date from about 3.9 billion years ago, obliterating any older craters. Each crater was formed by the impact of a body about 20 to 30 times smaller than the crater itself. Parts of the Earth would once have looked similar

average rate at which cratering has affected the Moon ever since. The Earth must have been exposed to the same flux of impactors as its satellite, and there are good reasons for believing that this is also a reasonable approximation for Mercury, Venus, and Mars. Counting craters is thus the best way we have to estimate ages on planetary surfaces. Even if the absolute age is in doubt, it is usually safe to assume that a surface with a lower crater density is younger than one with a higher crater density.

At present, the Earth is hit annually by about 10,000 meteorites greater than 1 kilogram, but most of those are too small to survive passage through the atmosphere, where they are heated and worn away by 'friction'. The yearly supply of 1,000-kilogram meteorites is only about 10, and the average interval between impacts by 150-metre-diameter objects (which would produce a crater some 2 kilometres across) is about 5,000 years. Impactors about 1 kilometre in diameter arrive at random about once every 200,000 years, boring through the atmosphere as if it were not there, hitting the ground with undiminished speed, and forming a crater perhaps 20 kilometres across. Larger, and more devastating, impacts are even less frequent.

Collisions affect each body in the Solar System, but craters survive only where there is a solid surface and insufficient other activity to erase the record. Observers were fortunate to discover a string of fragments of a tidally disrupted comet shortly before they collided with Jupiter in July 1994. Several of the collisions were witnessed, and each left a brown scar in the giant planet's atmosphere that lingered for several weeks, as did a scar found in July 2009 made by an unobserved single impact.

Planets as abodes of life

If Earth were not at a comfortable distance from the Sun, you would not be reading this book, because life may not have become established and we could not have evolved here even if it had. Scientists talk of a 'habitable zone' around every star, at a distance where the surface temperature on a planet would be neither too hot nor too cold for life. By analogy with Goldilocks' preference for baby bear's porridge (whose temperature was 'just right'), the habitable zone is sometimes called the 'Goldilocks zone'. In this context, 'habitable' means somewhere that could sustain life of any kind, even just simple microbes. It does not imply that the environment would be inhabitable by humans.

Because our kind of life requires water, the habitable zone is usually equated with the distance from a star at which the surface temperature of a planet would enable water to exist in liquid state. The density and composition of a planet's atmosphere influences the surface temperature, but the main control is heat received from the star. The habitable zone around the Sun is estimated to extend from about 0.95 to about 1.5 AU. These estimates put Venus (0.72 AU) well inward of the inner edge of the habitable zone, and Mars (1.52 AU) at its outer fringe. The Sun's output has probably increased slightly since the planets were formed, nudging the habitable zone outwards over time, so Mars would seem to be a poor candidate for life, but it is not a hopeless case.

The idea of a habitable zone defined by planetary surface temperature has been criticized as too narrow. There are circumstances where heat generated within a planet may provide an environmental niche suitable for life, although the surface may seem inhospitable. At the right height above its surface (about 50-60 km), the clouds of Venus are cool enough for liquid water (actually sulfuric acid) droplets, and there is a chance that perpetually airborne microbes could live there. Even on Earth, we know of 'extremophile' organisms living below 0 °C or above 100 °C. Thus, even if all life is, like that on Earth, based on chains of carbon molecules and dependent on water as a solvent, there are several places in the Solar System where it *could* exist (though only one, the Earth, where it is *known* to exist) and at least many millions of habitable places elsewhere in the galaxy. I shall return to that theme near the end of the book.

Space exploration

Telescopes are very useful, for example to measure the temperature and composition of a planet's surface and atmosphere. The polar ice caps on Mars were correctly identified by William Herschel as long ago as 1781. Jupiter is sufficiently large and close that storms among its clouds can be monitored even with fairly modest telescopes. However, this book would be duller and more speculative were it not for half a century of space exploration when space probes from Earth have visited every planet in the Solar System. Soviet probes reached the Moon in 1959, and twelve American astronauts walked on its surface between 1969 and 1972. Unmanned American (NASA) and Soviet probes flew past Venus and Mars in the 1960s, and achieved orbit and soft landings during the 1970s. The first Jupiter and Saturn fly-bys were in the 1970s, and the other giant planets were visited in the 1980s. Since 1990, the terrestrial planets have been explored by increasingly capable orbiters, robotic rovers have crawled across the Martian surface, and complex orbital tours of both Jupiter and Saturn have been achieved.

The most famous missions include *Vikings 1* and *2* that landed on Mars in 1976; *Magellan* that mapped the surface of Venus by radar 1990–4; *Voyagers 1* and *2* that flew past the giant planets between 1979 and 1989; *Galileo* that orbited Jupiter between 1995 and 2003; *Cassini* that began a thirteen-year orbital tour of Saturn in 2004 and sent a probe named Huygens to the surface of Titan in 2005; *MESSENGER* that orbited Mercury 2011-2015; *New Horizons* that flew past Jupiter in 2007, Pluto in 2015 and a small Kuiper belt object named Arrokoth in 2019; and *Rosetta* that orbited a comet 2014-2016 and deployed a lander.

Highlights in the years ahead include return to Earth of samples collected from Mars, asteroids, and comets, and a renewal of human presence on the Moon. The USA and Russia are no longer the only space powers. The European Space Agency has gone solo to Mars and Venus, to Saturn jointly with NASA, and is on its way to Mercury with Japan. The Japanese have sent probes to the Moon and to asteroids, and China and India have each reached the Moon. Scientifically, there has been much cooperation (and most probes carry instruments contributed by several nations), but it is undeniable that there is also national pride at stake, together with long-term strategic and commercial interests.

Chapter 2
Rocky planets

Here, I will discuss the planet that we live on and other bodies like it, namely the three other terrestrial planets Mercury, Venus, and Mars, and also the Moon. To the astronomers of the IAU, the Moon is just a satellite, but its composition and internal structure place it among the terrestrial planets from the perspective of a geologist or geophysicist. Figure 3 shows these five at the same scale, and Table 3 lists some relevant data. Within this group, Mercury and the Moon have effectively no atmosphere. Venus has only slightly lower size, mass, and density than the Earth, so gravity at its surface is only slightly less than on the Earth. However, its atmosphere is considerably denser. Mars is larger than Mercury but less dense. These two effects offset each other so that their surface gravities are very similar, but being colder, Mars has been able to hold on to a thin but respectable atmosphere. The Moon has the lowest surface gravity of all – about one-sixth of the Earth's – which is why Moon-walkers bound around so strangely. Mean surface temperatures obscure wide variations with latitude and, in some cases, between day and night. For example, the hottest daytime temperature on Mercury exceeds 400 °C, whereas at dawn after a long Mercurian night the temperature is below −180 °C.

Table 3 Basic data for the terrestrial planets

	Mass / 10^{24} kg	Polar diameter / km	Density / 10^3 kg m^{-3}	Surface gravity / m s^{-2}	Atmospheric pressure/ bars	Mean surface temperature
Mercury	0.330	4880	5.43	3.7	10^{-15}	170 °C
Venus	4.87	12104	5.20	8.9	92	480 °C
Earth	5.97	12714	5.51	9.8	1	15 °C
Moon	0.074	3476	3.34	1.6	2×10^{-14}	1°C
Mars	0.642	6750	3.93	3.7	0.0063	–50 °C

3. Top: from left to right, Mercury, Venus, Earth, Moon, and Mars, shown at the same scale. Bottom: the much larger giant planets Jupiter, Saturn, Uranus, and Neptune, with the terrestrial planets inserted to the same scale

Cores

Terrestrial planets are distinguished by having rocky exteriors, made largely of silicate minerals. However, their densities are too great for them to be rocky throughout, and it is believed that each has an iron-rich core at its centre. No planet's core can be seen or sampled directly, but there are several independent lines of evidence. Density is one, showing that the interior must be denser than rock even allowing for internal compression at high pressure, and analyses of the trajectories of orbiting spacecraft show that density increases symmetrically about each planet's centre. Chemical models for what is likely to happen inside a rocky planet suggest that there is insufficient oxygen for all the iron to be oxidized and bound up in silicate minerals. Thus, if the interior had ever been molten this would have allowed metallic iron, which is denser than rock, to sink towards the centre. This is an example of a processes called differentiation.

The outer parts of the iron-rich cores of the Earth and Mercury must be molten today, because those two planets have strong magnetic fields, apparently generated by dynamo motion in an electrically conducting fluid. For such a small planet, Mercury's density is very high, so its core must be exceptionally large, occupying about 40% of its volume and accounting for nearly 75% of its mass. Magnetic fields are not being generated inside Venus, the Moon, and Mars, so their cores are probably entirely solid.

In the case of the Earth, we have additional evidence about the core from studying how seismic waves, which are vibrations triggered by earthquakes (or underground nuclear tests!), travel through the planet. This confirms a solid inner core 1,215 kilometres in radius and a fluid outer core 3,470 kilometres in radius. Both seem to be mainly iron alloyed with 5%–10% nickel, but density arguments require something less dense than iron too, making up 6%–10% of the outer core and 2%–5% of the inner core. The likeliest explanation is some mix of oxygen, silicon, and sulfur.

In total, the Earth's core occupies about 16% of the planet's volume. Comparable values for Venus and Mars, which are estimates based largely on their average densities, are about 12% and 9%, respectively. There are some very limited seismic data from the Moon (from the Apollo programme), hinting at a relatively small core between about 220 and 450 kilometres in radius (less than 4% of the Moon's total volume). About 1 in every 20 meteorites is made of an alloy of iron with 4.5%–18% nickel, corresponding to the cores of planetesimals from the asteroid belt that had differentiated internally before being broken up by collisions.

Mantles and crusts

The silicate part of a terrestrial planet surrounding its core is called the mantle. This makes up the majority of the total volume of each terrestrial planet, and most of their mass except for

Mercury. The crust is a relatively minor unit overlying the mantle. It is also of silicate, though slightly different in composition to the mantle.

A planet's present mantle evolved from the molten rock that would have covered the globe after the final giant impact collision, known to geologists as a 'magma ocean'. While a magma ocean cools, its surface must radiate heat to space, causing it to chill to a solid skin. However, this skin would continually be broken and stirred back in, thanks to turbulence below and impacts from above. The magma ocean would continue to cool, but, unlike the freezing of a ball of water, there is no discrete temperature at which the whole of it would become solid. The nature of molten silicate material is such that minerals of various compositions crystallize out at different temperatures and pressures. Planetary scientists are unsure of the extent to which magma oceans crystallized in layers, or whether minerals denser than the melt were able to sink while less-dense minerals were able to rise, perhaps sticking together to form massive 'rock bergs' that could force their way up more effectively.

Aggregations of this flotation material, chemically different to the underlying magma ocean, formed the earliest true crust on the Moon, where it survives today as the lunar highlands (the pale areas on the Moon's face). On the larger terrestrial planets, the nature of the oldest crust has not been determined, partly because it has largely been replaced (or, at least, covered over) by later kinds of crust. To see how this might happen, we have to turn our attention back to the mantle. As a young planet cools, eventually its mantle becomes fully solid. Two important characteristics of silicate materials now become relevant. The first is that sufficiently hot solids are neither completely immobile nor undeformable. Hot rock in a planet's interior is capable of flowing at speeds of a few centimetres per year (the rate at which your fingernails grow), in much the same way as a block of pitch will deform over time. Within a solid mantle, motion will occur at a slow but geologically effective pace if there are any forces capable of driving it. Inside a planet, the necessary impetus comes from heat. Hotter mantle from deep down will be slightly less dense than the cooler mantle above, so there is a tendency to swap places. Motion of this sort is called convection, and is what you can observe in a pan of soup heated on a hob, except that within a planet it is much slower 'solid state convection'.

Imagine a streamer or 'plume' of hot mantle welling upwards and displacing colder mantle downwards. As it gets closer to the surface, the pressure experienced by the rising mantle decreases, which brings into play the second relevant characteristic. As pressure falls, silicates begin to melt. The process is called 'partial melting', because only part of the solid melts, and the magma that forms is slightly richer in silica than the solid from which it was extracted. The resulting magma is also less dense than the solid, so buoyancy forces will squeeze it upwards towards the surface, especially if there are pathways where the overlying rock is under tension or fractured. Unless it stalls below ground as an intrusion, the magma will be erupted through volcanoes.

Rock formed in this way is described as igneous, and crust produced by igneous activity can replace the original crust of a planet by infiltration or burial. The dark patches on the Moon, the lunar 'maria', are low-lying regions where the paler primary crust has been buried by lava flows produced in this way. The present-day crust of the Earth results from partial melting of the mantle to make oceanic crust, and from melting and recycling of many generations of oceanic crust to make continental crust. The Earth's oceanic crust is 6–11 kilometres thick, whereas continental crust varies from about 25 kilometres in thin, stretched regions to 90 kilometres below major mountain ranges. In total, crust occupies only about 1% of the Earth's total volume. The Moon's crust averages about 70 kilometres in thickness (13% of the Moon's volume), ranging from >100 kilometres in some highland regions to <20 kilometres below some major impact basins.

In summary, crust is chemically related to the underlying mantle, but differs in ways depending on how it was extracted from it. Crust is lower in density and its average composition is richer in silica than the mantle. Crust is more varied than mantle, and includes rock that has reacted chemically with any atmosphere or liquid water, and that has been broken apart or dissolved, transported (by gravity, wind, water, or ice), and deposited elsewhere. Such deposits constitute sedimentary rock. Burial, deformation, and heating can cause sedimentary or igneous rock to recrystallize, in which case it is known as metamorphic rock.

Internal heat

Planets are hot inside partly because of heat left over from their accretion. For a bigger planet, the fraction of this 'primordial heat' remaining today is greater. This is because heat content is related to planetary volume, which depends on the cube of the radius, whereas heat leakage is limited by surface area, which depends only on the square of the radius.

Heat is also generated inside a planet, by decay of radioactive isotopes. There are many of these, but only four whose decay produces significant heating: potassium-40, uranium-238, uranium-235, and thorium-232. Because of their geochemical affinities, these elements are more abundant in crustal rocks than in the mantle. In the Earth, approximately the same amount of heat is generated radiogenially (that is, by radioactive decay) in the crust as in the whole of the volumetrically much larger mantle.

A terrestrial planet's total content of heat-producing elements depends on its mass (and hence its volume). Just like primordial heat, radiogenic heat is retained more effectively in larger planets. In the case of the Earth, about half the heat leaking to the surface today is primordial, and almost all the rest is radiogenic.

Lithospheres

The transition in properties from cold and rigid to warm and convective generally occurs at some depth below the boundary between crust and mantle. Thus the crust and the uppermost mantle constitute a single mechanical layer, forming a rigid outer shell. This shell is called the 'lithosphere', using the Greek word *lithos* ('rock') to indicate that the layer has the mechanical properties of everyday rock. Below the lithosphere is where the mantle, although rocky in composition, is hot enough and weak enough to convect. This zone is sometimes called the asthenosphere (using Greek *a-sthenos* meaning 'without strength').

The Earth's lithosphere is about 100 kilometres thick and is fractured into a number of plates, which are able to be shunted around thanks to the special weakness of the underlying asthenosphere. As part of a process known as 'plate tectonics', new lithosphere is created where plates pull apart (usually hidden from view below the ocean) and destroyed where one plate is drawn below another, at subduction zones marked by trenches on the ocean floor. The grinding of one plate against its neighbour is the cause of most earthquakes. If anyone tells you that the Earth's plates are 'crust sliding over the mantle', they are wrong, repeating a persistent fallacy that appears in many school textbooks and examination syllabuses. The truth is that a plate consists of crust and the conjoined rigid uppermost mantle, which together slide across the deeper, less rigid, asthenospheric mantle.

The lithosphere, being brittle, is the layer where faults can occur, as one mass of rock grinds past another. Faults are common on the Earth, especially in the zones where two plates meet, and can be identified on other planets too (Figure 4).

Plate tectonics seems to be unique to the Earth. Greater lithospheric thickness in the more easily cooled smaller bodies of Mercury, the Moon, and Mars undoubtedly contribute to this, but

4. A 500-kilometre-wide view of part of Mercury. Solar illumination is from the right. Shadow picks out a kilometre-high scarp with the shape of an open letter M on its side. This is an ancient thrust fault named Beagle Rupes, marking where the terrain on the right (east) has been pushed westwards over the terrain on the left (west). Some of the craters are older and others younger than this fault

a more important factor is that for plates to be mobile, the top of the asthenosphere needs to be especially weak. Within the Earth, this is accomplished because of a small amount of water within the rock, which weakens it and encourages the formation of a small amount of melt that lubricates grain boundaries. Venus has lost its water, so its asthenosphere is dry and lithospheric plates cannot slide freely across it.

A planetary asthenosphere that is dry, or very deep, is manifested principally by two effects at the surface. One is the height of mountains and depth of basins. If these are too great, the asthenosphere will flow and flex the overlying lithosphere, thereby reducing the topographic contrast until it is small enough to be supported by the strength of the lithosphere alone. The second is the pattern of fracturing caused by large impacts. An impactor several tens of kilometres in diameter arrives with sufficient force for the resulting crater-forming shockwaves to disrupt the lithosphere, and the crater will take the form of a basin marked by concentric rings of fractures. In a thinner lithosphere, the rings tend to be closer together, so these multi-ringed impact basins can be used to estimate the depth to the asthenosphere at the time of their formation. As a planet slowly cools, its lithosphere becomes gradually thicker.

Volcanism

Magma, the name given to molten rock before it erupts, can be generated inside a planet essentially by three different causes. Direct application of heat is only one of these, and is often the least important: slow build-up of heat trapped below a planet's lithosphere may account for some episodes of widespread volcanism, and strongly varying tidal stresses inside a planetary body work against internal friction, leading to 'tidal heating'. Alternatively, decreasing pressure in an upwelling zone in the mantle can cause partial melting (for example, leading to the generation of Earth's oceanic crust). Also, it is possible that *sudden* reduction in pressure, as must happen to the mantle below where a major impact basin is excavated, could trigger a melting event. The third mechanism is to introduce water into the mantle or lower crust. Water reduces the temperature at which silicates begin to melt. The Earth has chains of volcanoes above subduction zones, because water that has been dragged down within the rocks of the subducting plate escapes upwards into the base of the over-riding plate. Here conditions are not hot enough for melting when dry, but partial melting will begin as soon as water is introduced even though there has been no rise in temperature.

The Moon

People began to speculate about volcanism on the Moon almost as soon as craters were seen through telescopes. They were on a false trail because, as we are now quite certain, almost all the craters on the Moon were made by impacts. In fact, the major volcanic areas of the Moon are the dark patches that were once thought to be dried-up sea beds. That is not the case, though they still bear the name 'mare' (pronounced mah-ray), which is Latin for 'sea'. The plural is 'maria' (pronounced mah-ri-a). These cover about 17% of the Moon's surface, mostly on the near side, which is the hemisphere that permanently faces the Earth. Here, lava with a composition similar to terrestrial basalt has flooded the major multi-ringed impact basins.

Specific vents where the mare basalts were erupted are hard to identify (Figure 5). Clearly, they did not take the form of conical volcanoes. Most likely, they were fissures through which fountains of incandescent molten lava were expelled by the force of expanding volcanic gas to heights well in excess of a kilometre. On falling to the ground, the lava was still hot enough to spread out, flowing downhill for hundreds of kilometres. Most of the fissure vents sealed themselves over as their rate of eruption waned, or were buried by later eruptions.

Four of the six Apollo Moon landings (1969–72) were on maria, which are flatter and so safer places to land than the highlands. Samples of mare basalt brought back for analysis can be dated with high precision by measuring radioactive decay products within them (radiometric dating), and the Apollo samples show a range of mare ages from 3.9 to 3.1 billion years. This long duration of volcanism put paid to the simplest volcanic explanation for the maria, which was that the volcanism was a direct product of the basin-forming impacts. Furthermore, work since the year 2000 has identified some patches of mare bearing sufficiently few superimposed impact craters that they must be younger than

5. A 200-kilometre-wide view of the south-east edge of Mare Imbrium. The rugged terrain on the right is highland crust uplifted in part of the basin rim. The darker, smoother area in the upper left is mare basalts that have flooded the lower-lying ground. A 1-kilometre-wide trench named Hadley Rille winds from south to north across the centre of the view, and this is believed to be a pathway along which lava flowed from a source largely obscured by shadow. *Apollo 15* landed close to Hadley Rille, near the middle of the image

about 1.2 billion years. Conversely, in 2007 a fragment of lunar material found on Earth as meteorite (having previously been flung off the Moon as ejecta from an impact crater) was found to contain fragments of basalt dated at 4.35 billion years, half a billion years before the late heavy bombardment ended. No maria of such great age remain visible, having been buried by ejecta from subsequent basin-forming impacts. So now we know that lunar volcanism began early as well as finishing late.

Mercury

Mercury is much less well known than the Moon. Under half of it was imaged by NASA's *Mariner 10* in 1974–5, and the planet then remained unvisited until NASA's MESSENGER probe began a series of fly-bys in 2008 and orbited 2011–2015. This revealed details sufficient to overcome most people's scepticism over the extent of volcanism on Mercury. For example, in Figure 4 the smooth terrain in the lower right and filling the 120-kilometre-diameter basin just above right of centre is now accepted as volcanic. Previous doubts were heightened by the fact that Mercury lacks the contrast in albedo (reflected brightness) between paler highlands and darker lavas that makes the maria so obvious on the Moon. This seems to be because the minerals making up Mercury's lavas contain much less iron than is typical of lunar (and terrestrial) basalts. Plains formed by lava probably constitute the majority of Mercury's surface. Some of them are old enough to date back to the era of the late heavy bombardment and are densely cratered, others are younger and have fewer craters superimposed on them.

MESSENGER imaged more than a hundred volcanic vents surrounded by bright haloes interpreted as explosive eruption deposits. For a planet so close to the Sun, Mercury's surface is unexpectedly rich in volatile elements such as sulfur, sodium and chlorine, some of which probably provided the gases needed to drive the explosive eruptions. There are also patches referred to as 'hollows' where the top 10 to 20 metres of surface material seems to have been stripped away to space; we don't know how, but it surely requires the lost material to have been volatile. We will find out more when the European Space Agency's BepiColombo mission starts providing data from orbit about Mercury in 2026. At present, it is safe to say that extensive areas of lava were emplaced during a period covering at least 4 to 3 billion years ago, and possibly extending to within the past billion years. Such a long duration of volcanic activity on Mercury was not anticipated, and may result from the same mysterious heat source that keeps part of its core molten.

Venus

Venus is much bigger than Mercury. Its size and mass would suggest almost as much radiogenic heat production as the Earth, and hence a similar level of volcanic activity. However, because Venus lacks plate tectonics, its volcanism operates rather differently.

Venus has a dense, permanently cloudy atmosphere, which kept the surface a total mystery until it became possible to study it by means of radar. Figure 6 shows a radar image of part of Venus obtained by NASA's *Magellan* probe, which mapped almost the entire planet between 1990 and 1994. Radar images are assembled by complex analysis of the echoes bounced back in response to a continual string of radar pulses beamed at the surface. For most purposes, you can treat radar images like the black-and-white optical images that they resemble, though in fact the brightness of each feature is controlled mainly by how rough the local surface is, rather than its albedo in visible light.

Figure 6 typifies much of Venus. It shows numerous individual lava flows – some rougher (brighter) and some smoother (darker) – that flowed from west to east across the image. The lobate pattern of the individual flows closely resembles that of lava flows on Earth and Mars, but which is hard to discern on the Moon and Mercury where the flow margins have become degraded by impacts.

As well as lava flows covering about half its surface, Venus has many clearly identifiable volcanoes. Figure 7 shows an example. In the background is a 5-kilometre-high volcano with gently sloping flanks of the kind known on Earth as a 'shield volcano' that results from repeated eruption of basalt through a single vent. Some individual lava flows can be made out on the flanks. No one is sure how long ago this volcano and others like it last erupted. There have been intriguing hints but no proof of recent or present-day

6. A 500-kilometre-wide *Magellan* image of part of Venus. The area is mostly lava, fed from a source 300 kilometres west of the image, but in the south-east corner is some rugged terrain representing the oldest surviving crust on Venus. Running from north to south in the west of the image is a mountain belt of ridged and fractured terrain that is breached by the lava flows

activity on volcanoes such as this, and they are rather too small for reliable crater-counting statistics. This particular volcano is built upon an older terrain unit that is smoother, except for numerous fractures upon it. The impact crater in the foreground is probably unrelated to radar-bright lava flows immediately to its left.

Circular or elliptical patterns of concentric fractures termed 'coronae', of which more than 300 have been identified on Venus, are not thought to share a common origin with the multi-ringed impact basins of the Moon and Mercury. They can be anything from 200 metres to over 2,000 kilometres across, and are usually associated with some form of volcanism. Probably each corona marks a site where an upwelling plume in the asthenospheric

7. Computer-generated three-dimensional perspective view showing the volcano Maat Mons on Venus. This was made by draping a radar image over a model of the topography obtained by radar altimetry. Vertical scale is exaggerated tenfold. Both sets of data were collected by the *Magellan* orbiter. The impact crater in the right foreground is 23 kilometres across

mantle impinged on the base of the lithosphere. Coronae where the plume is still present are uplifted as very broad domes, whereas older ones, no longer supported by a mantle plume, have sagged. The sagging, in particular, explains the concentric fractures.

Impact craters on Venus are more common than on the Earth, but considerably less abundant than on the Moon and Mercury (you will not find any in Figure 6). There are two factors at play here. Craters less than about 3 kilometres across are entirely absent from Venus, because its dense atmosphere shields the surface from small impactors. However, larger craters are formed by objects carrying too much energy to be affected by the atmosphere. Their lack of abundance has to be explained by the young age of the surface, which works out on average to be between about 500 million and 700 million years. There are no large tracts of terrain that seem to be very much older or very much younger than the global average.

The standard explanation for this in the 1990s was that pretty much the whole planet was resurfaced in an orgy of volcanism that began 500–700 million years ago, and lasted no more than a few tens of millions of years. This would be consistent with Venus's lack of plate tectonics leading to most heat from the deeper mantle being trapped below the lithospheric lid, until much of the uppermost asthenosphere had melted. Eventually, the cold, dense lithosphere would founder, and the buoyant magma from below would erupt. Something similar could have occurred half a dozen times since Venus was formed, and maybe it could happen again within the next 100 million years.

This model calling for catastrophic global volcanism has recently been challenged, on the grounds that the cratering statistics do not rule out a more gradual process. Progressively smaller areas could have been resurfaced by lavas at random intervals throughout the past half billion years.

Earth

On Earth, volcanism and plate tectonics combine to regulate the internal heat budget and thus prevent major asthenospheric temperature excursions of the kind postulated for Venus. Only about one-third of the heat generated below the lithosphere leaks out by conduction. Most of the heat is conveyed to the top of the lithosphere by eruption at mid-ocean ridges (where new material is added to diverging plates) and, to a smaller extent, by eruptions at volcanoes above subduction zones and at various 'hot spots' above mantle plumes. The asthenosphere is cooled by the reincorporation into it of the old, cold parts of lithospheric plates at subduction zones.

The closest we get to a Venus-like volcanic catastrophe is when, every few tens of million years, a region maybe a thousand kilometres across is buried by the eruption of up to ten cubic kilometres of basalt lava. This is known as a 'flood basalt'. The 'Deccan Traps' of north-west India (66 million years), the Brito-Arctic flood basalts (Greenland and the north-western British Isles, 57 million years), and the Colombia River flood basalts (north-western USA, 16 million years) are among the better-known examples. These major but rare events could be capable of injecting so much volcanic gas, notably sulfur dioxide, plus fine fragments of volcanic rock known as 'ash', into the atmosphere that global climate could be severely affected. Figure 8 shows an example of lava flows on the Earth, for comparison with the images from other planets.

The way in which volcanism on the Earth probably differs most from the other planets is that expansion of gas within rising

8. A 70-kilometre-wide view from space showing the 'Craters of the Moon' lava field in Idaho, USA. The source of the flows was a series of fissures near the edge of the rugged highlands in the north-west. Compare the lobate form of the lava flows with the flows on Venus in Figure 6

magma tends to make a substantial proportion of eruptions explosive in nature. This is for two reasons. The first is that recycled water, carbon dioxide, and sulfur dioxide escaping upwards above subduction zones adds greatly to the leakage from the deeper interior of primordial gases, so the Earth has more gas available to drive eruptions explosively. The second is that the existence of continental crust facilitates generation of magma with a higher silica content than basalt. These silica-rich magmas are more viscous than basalt, so they fragment more easily. Classic 'picture-book' steeply conical volcanoes such as Mount Fuji in Japan are rare except on Earth, because they are symptomatic of relatively silica-rich and partly explosive eruptions.

Mars

Compared to the Earth and Venus, Mars has relatively few volcanoes. However, their small numbers are compensated by large size. Major groupings of large basaltic shield volcanoes occur in the Tharsis region (much of which is included in Figure 9) and the Elysium region. Olympus Mons is the largest Tharsis volcano, measuring about 600 kilometres across its base and 24 kilometres from top to bottom, which makes it the largest volcano in the entire Solar System. There are two reasons why Mars has such big volcanoes. The first is that Mars is a 'one plate planet'. Its lithosphere is an intact shell (a single plate) effectively stationary with respect to the underlying mantle asthenosphere. Unlike the Earth, where plates drift around relative to mantle plumes so that plume-fed volcanoes are carried away and cut off from their magma supply after only a few million years, a mantle plume on Mars supplies magma to the same spot in the lithosphere for as long as the plume remains active. Olympus Mons may have begun to be constructed more than a billion years ago. We have no way to tell, because we can only date (by crater counting) what is exposed at the surface today, and can't see the older, buried, interior of the edifice. There are several overlapping calderas at

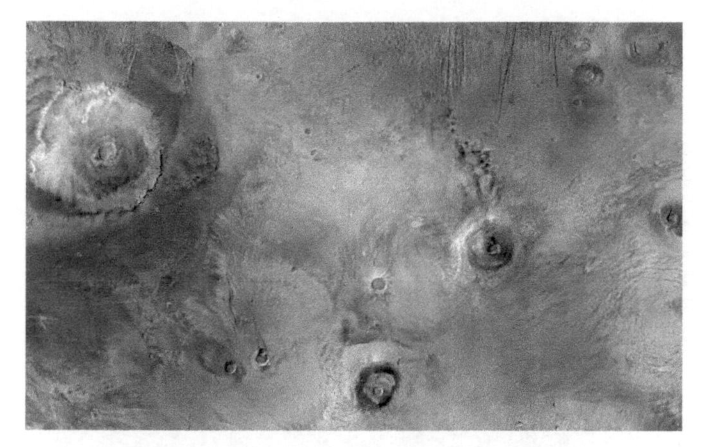

9. A 3,000-kilometre-wide mosaic of images showing several enormous shield volcanoes on Mars. On the left is Olympus Mons, the Solar System's largest volcano. On the right-hand edge is Tharsis Tholus, and running north-east from the centre of the southern edge, a line of three: Pavonis Mons, Ascraeus Mons, and Ceraunius Tholus

its summit, whose floors are dated at around 100 to 200 million years, but the youngest lava flows on the flanks appear to be only about 2 million years old, and it is likely that Olympus Mons will erupt again one day. Other volcanoes in the Tharsis region are definitely older, and are probably now extinct.

The second reason why Mars has such big volcanoes is 'because it can'. It has a cold, strong lithosphere, about twice as thick as the Earth's. If you transplanted Olympus Mons to the Earth or Venus, their relatively thin lithospheres would sag beneath the load, and the volcano would lose height.

High-resolution images reveal details of lava flows on the plains between the large volcanoes and in several other regions of Mars. However, there are also features regarded by some as volcanic that have aroused considerable controversy. Figure 10 shows a notable example.

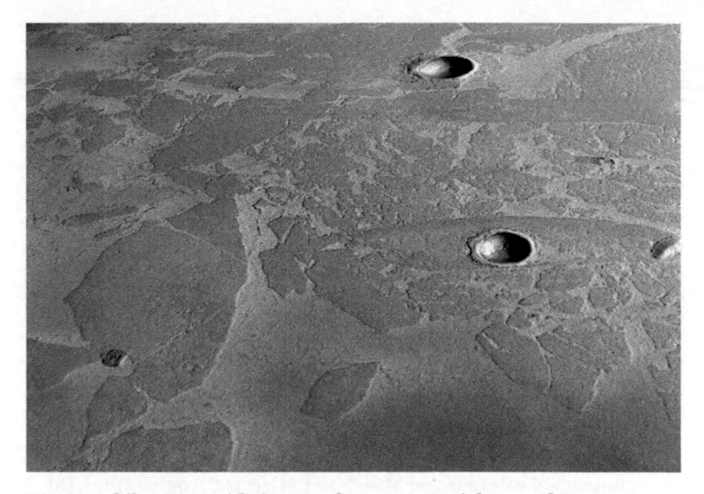

10. A 50-kilometre-wide image of a controversial area of Mars obtained by ESA's *Mars Express* orbiter. Some say the platy surface is a lava flow with a fractured cooling crust. Others see this as broken pack ice (now dust-covered) on the surface of a frozen sea. The two impact craters are older than the platy surface, and their rims were high enough to prevent their interiors from flooding. The craters are actually circular in outline, but are foreshortened in this oblique view

Over 30 fragments of impact ejecta from Mars have been collected on Earth as meteorites. They are all either basaltic lava or more coarsely crystalline intrusive equivalents, spanning a range of crystallization ages extending from 4.5 billion years to as young as 160 million years. We can infer that igneous rock makes up the bulk of the Martian crust at depth, even though large tracts of surface have a veneer of sediment of various kinds.

Surface processes

Regolith and space weathering

Volcanism is driven from inside a planet, but planetary landscapes can be sculpted just as much by processes that occur essentially at the surface. On a body that is airless and therefore unprotected from space, by far the dominant process acting directly on the surface is bombardment by meteorites and micrometeorites. Fragmented material ('ejecta') thrown out from craters blankets the surface to a depth of several metres, and sites where solid bedrock is visible at the surface are rare (Figure 11). The lunar soil, known as 'regolith', in which the Apollo astronauts left their footprints is composed of grains mostly a fraction of a millimetre in size, comprising fragments of crystal, tiny bits of rock, and glassy spherules that are frozen droplets of melt generated by the heat of the impact. Regolith is continually rearranged on a variety of scales by excavation of craters and dispersal of ejecta, a process known as 'impact gardening'. On Mercury, where impact speeds are faster, regolith grain-size is expected to be about one-third that of lunar regolith.

11. Telephoto view looking across Hadley Rille, taken by Apollo astronaut Dave Scott. The 2-metre-thick horizontal layer running across from the left is a rare example of bedrock (probably a lava flow), here exposed on a steep slope. Everywhere else is covered by regolith ranging in size from boulders down to dust

If there is no atmosphere, solar ultraviolet light can reach the surface, where it may, over time, break chemical bonds. Micrometeorite impacts and (if there is no magnetic field) charged particles from the solar wind can also affect surface chemistry, so airless bodies experience a suite of processes, collectively described as 'space weathering', that slowly alter the composition of the surface. For example, the bonds linking iron to oxygen atoms can be broken, allowing oxygen to escape and leaving submicroscopic grains of pure metal, called 'nanophase iron'.

When a planet has an atmosphere, only the largest, and infrequent, impactors can reach the surface at high speed. For example, in the Earth's atmosphere, stony asteroids less than about 150 metres in size are likely to be sheared apart. The resulting fragments are small enough to be slowed down by friction, so by the time they reach the ground they have lost almost all their initial velocity, and do not form craters. Meteoritic dust, which is mostly micrometeorites but also fragments frictionally ablated from larger meteorites, settles to the ground at an average accumulation rate of 0.1–1 mm per million years. This dust makes such a tiny contribution to the total rate of sedimentation that it is totally swamped by other sediment, except on deep ocean floor far from land.

Erosion and transport

Impact gardening aside, processes that can wear away rock and transport the resulting fragments are wind, flowing water, and moving ice (glaciers). Water can also dissolve rocks, during chemical weathering. Elements carried in solution by water may reappear elsewhere, being precipitated in new minerals. This applies especially to salt deposits, and also to many forms of carbonate rocks. However, on Earth most limestone (calcium carbonate) is formed from fragments of the shells of marine organisms, demonstrating an important biological step in turning dissolved carbonate (or dissolved carbon dioxide gas) into solid material that can become rock.

The dust storms on Mars are famous, having been first noted by telescope in 1809. At perihelion, when Mars receives 40% more solar energy than at aphelion, winds in excess of 20 metres per second can lift so much dust high into the sky that most of the surface is obscured for several weeks. Sometimes little can be seen poking through other than the summit of Olympus Mons. Because of the clouds that often congregate there, this often looks white, hence its former name of Nyx Olympica (Snows of Olympus), which was revised when images from spacecraft showed what was really going on.

12. Some sand dunes, needing only camels or a palm tree for scale. In fact, this picture was taken by NASA's *Opportunity* rover on the surface of Mars, looking obliquely down from the rim of a crater onto a field of dunes on the crater floor. The visible area is about 100 metres across

Many signs of the action of wind on Mars can be seen from orbit or on the ground (Figure 12) in the form of sand dunes and also smaller wind-ripples in the surface dust. Some of the dunes on Mars are being actively sculpted by the wind, but others have probably not changed their form for millions of years. Wind-blown sand is a powerful agent of erosion on Mars. The low atmospheric density means that a wind capable of transporting sand grains has to be blowing much faster than on Earth, and some exposed layers of rock have become curiously sculpted by abrasion.

Venus has a much denser atmosphere than Earth, with 92 times greater ground-level atmospheric pressure. Even sluggish winds can shift sand particles around, and Venus has several fields of sand dunes. However, here when a wind-blown grain strikes an exposure of bedrock its erosive power is limited, partly because of the slow speed and cushioning of the blow by the dense air, and partly because the high surface temperature of 480°C makes material deform plastically rather than abrading in a brittle fashion.

For Earthlings, flowing water is usually the most familiar agent for transporting sediment – in a river, or as waves on a beach. In the Solar System, nowhere other than Earth currently has surface conditions allowing liquid water to be stable. Venus is far too hot, and although the noontime temperature on Mars can creep well above 0 °C, its atmosphere is so tenuous that ice at the surface turns directly to vapour rather than melting. However, there is abundant evidence that water once flowed in prodigious volumes across the surface of Mars (Figure 13). Mars has suffered extremes of climate at least the equal of Earth's, and billions of years ago its atmosphere was sufficiently dense and wet to permit rainfall and catastrophic flooding. The largest canyon system in the Solar System, Valles Marineris ('the Valleys of the Mariner', named after being discovered on images returned by the probe *Mariner 9* in 1971), is a 4,000-kilometre-long rift system initiated by fracturing of the crust, but widened by erosion when water flowed through it.

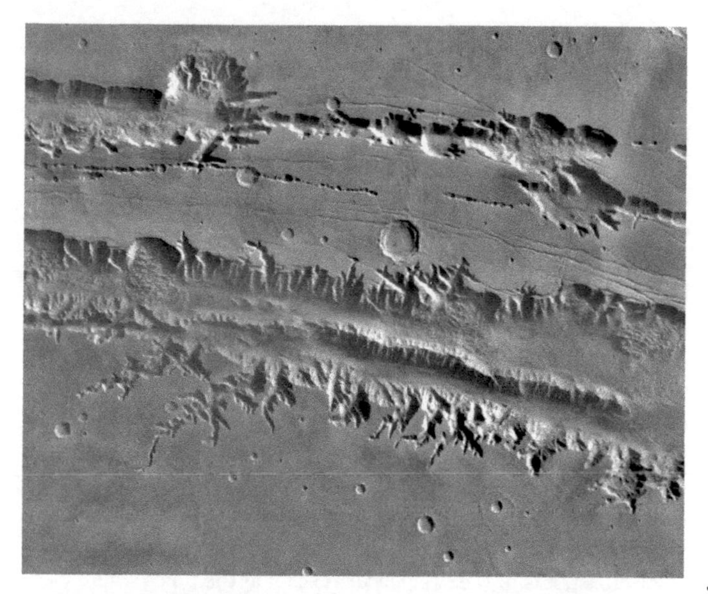

13. A series of east–west fractures attests to the tectonic origin of Mars's Valles Marineris canyon complex, of which only a fraction is covered by this 800-kilometre-wide view. Note the winding and deeply incised channels feeding into it from the south, which show the role played by flowing water in widening the main canyon

At its deepest point, the floor is 7 kilometres below the rim (Earth's Grand Canyon in Arizona is only 2 kilometres deep), and it is so wide that if you stood on one rim, the opposite side would be out of sight beyond the horizon.

Despite its vastness, Valles Marineris was not recognized by pre-Space Age telescopic observers. The notorious 'canals' of Mars mapped by the Italian Giovanni Schiaparelli in 1877, and subsequently championed by the American Percival Lowell who until his death in 1916 thought they were giant works of engineering by intelligent Martians, are illusory. They bear no relation to any of the many genuine channels on Mars. Of these, examples fed by a branching network of tributaries (including many much longer versions of those shown in Figure 13) are likely to have been supplied by rainfall. The water flowing along others probably leaked out of the ground, possibly when permafrost melted. The streamlined shapes of the 'islands' where channels debouch onto the plains show them to have been scoured by catastrophic floods. Robotic landers (*Viking 1* in 1976 and *Mars Pathfinder* in 1997) that touched down in such places found an abundance of rocks dumped there by floodwaters.

All the major valleys on Mars have numerous impact craters superimposed upon them, so clearly they must be ancient, having last flowed well over a billion years ago. Since then, many have suffered landslips from their walls, and their floors now bear trains of sand dunes formed by the cold wind howling along their length. In the 1970s and 1980s, most scientists would have told you that although Mars experienced at least one wet epoch in its distant past, it is now extremely dry except at the poles where there are small caps of water-ice. Imagine everyone's surprise when in 1999 a high-resolution imager called the Mars Orbiter Camera began to reveal gullies only a few metres wide and a few hundred metres long on steep slopes in several places on Mars. The lack of superimposed craters and the observation that in many cases debris fans at their downstream ends had begun to bury sand dunes showed they must be young features, but how young? Proof was not long in coming that some are still active today, when repeat images began to reveal changes (Figure 14).

The debate has now moved on from questioning the age of the youngest gullies, and now focuses on how they are cut. One theory is that water is responsible. There could be reservoirs of liquid groundwater under pressure in the Martian subsoil. Where a slope, such as the crater wall in Figure 14, cuts below the water table, a barrier of ice within the soil ('permafrost') would normally prevent its escape. However, if the barrier were temporarily to give way, water could come gushing out. The liquid would not be stable – it

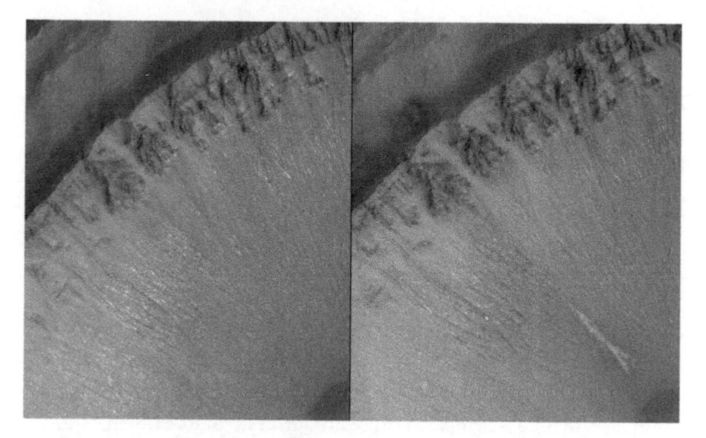

14. Two views of the same 1.5-kilometre-wide area covering the inner wall of a 6-kilometre-diameter Martian crater, recorded in August 1999 (left) and September 2005 (right). The rim cuts across the top left, and the floor is towards the lower right. There are many gullies eroded into the slope of inner wall, one of which seems to have flowed between these two dates, carrying some pale debris onto the lower slope

would be both boiling and freezing as it flowed – but it could traverse the length of one of these gullies before completely evaporating. Sceptics argue that liquid flow is not necessary to carve the gullies, and that they can be explained as a result of dry rock avalanches.

Some Mars scientists see evidence for glaciers, especially at the eroded edges of highland plateaus. There is no ice exposed at the surface today (except that at the poles), but the rock-strewn ground revealed in high-resolution images from orbit could be debris covering (and insulating) underlying ice. Ground-penetrating radar data obtained from Mars orbit lend credence to this, which is one of the reasons why I am happier to accept the region shown in Figure 10 as a dust-covered frozen sea rather than as a lava flow.

Channels on the Moon such as Hadley Rille (Figure 5) were lava pathways and were certainly not cut by water, and the only lunar water is small quantities of ice in the regolith near the poles. More than 200 sinuous channels have been mapped on Venus, one of which is 6,800 kilometres long. It is very unlikely that Venus experienced sufficiently extreme climate change for liquid water to have existed recently enough to erode these channels, so they too were probably cut by lava.

Naming surface features

I have used names of surface features on other planets several times already: Olympus Mons, Valles Marineris, Hadley Rille, and so on. Without such names, I would be reduced to referring to them as 'the biggest volcano on Mars', 'Mars's giant canyon system', and 'that big trench near where *Apollo 15* landed'. Less notable features would be even harder to describe, unless by a totally unmemorable coordinate system.

But nobody lives there, so who allocates names and how official are they? When astronomers first started to draw maps through their telescopes, some were sufficiently independently minded to invent their own names, often regardless of any previous work. An early task of the IAU (founded in 1919) was to sort out the mess, arrive at single official names for multiply named features, and to establish standards and conventions for allocating future names. This applied to the names of newly discovered bodies and also features on the surfaces of planetary bodies that it might become desirable to name or that would become visible thanks to improved imaging techniques. Originally, the latter meant simply bigger and better telescopes, and few founders of the IAU can have realized that they had established a means for supervising the nomenclature of features revealed by visiting spacecraft.

Some have castigated the IAU's handling of Pluto's reclassification, but I know of no one who thinks badly of the basis for the IAU-administered naming process. This is fair, non-political, and seeks to represent all the world's cultures – not necessarily on any single body, but balanced across the whole Solar System.

Building on what had already become common practice for lunar features, IAU nomenclature assigns a single unqualified name to craters, whereas most other features are given a name plus a Latin descriptor term that denotes what kind of a feature it is. Thus 'Olympus Mons' means 'Olympus Mountain', telling you immediately that this feature is a mountain named Olympus. Note that although no one doubts that Olympus Mons is a volcano, the descriptor term does not say this. Descriptor terms intentionally avoid *interpretation* (which may turn out to be wrong) and stick to *description*.

Common descriptor terms that you may meet are: chasma (a deep, elongated, steep-sided depression), fluctus (a flow-like feature), fossa (long, narrow, shallow depression), mensa (flat-topped prominence with cliff-like edges), planitia (low-lying plain), planum (high plain or plateau), rupes (scarp), and vallis (branching valley). On the Moon, there is also mare (plural maria) which translates as 'sea' but had become too entrenched to be replaced by a more apt term.

There are also themes for names on each planet. Lunar craters are named after famous deceased scientists, scholars, and artists, whereas maria take Latin terms describing various weather conditions. Mars is the only place other than the Moon with significant heritage of names from before the IAU became involved. These, with modern descriptor terms added, come from telescopic mapping by Giovanni Schiaparelli and Eugenios Antoniadi in the late 19th century, and mostly refer to broad regions such as Tharsis and Elysium. Each large valley is given the name for Mars in a different language, whereas small valleys are named after rivers on Earth. On Venus, almost all names are female: craters are named after famous historical women and most other features after goddesses. On Mercury, craters are named after dead artists, musicians, painters, and authors, whereas scarps (rupes) are named after scientific expeditions or the ships that carried them. Beagle Rupes (Figure 4) is named after *HMS Beagle* on which Charles Darwin voyaged while amassing the observations that inspired his theory of evolution.

Similar principles apply to names on asteroids and the satellites of other planets. For example, Jupiter's satellite Europa has craters named after Celtic gods and heroes, and most other features have names taken from the classical myth about Europa.

Atmospheres

After its birth, each terrestrial planet must have developed an atmosphere when internal gases leaked out via the magma ocean. These primitive atmospheres do not survive today, though the gases escaping from volcanoes show what they may have been like. The Moon and Mercury have too little gravity to hang on to a gas blanket, and the 'atmospheres' that you can sometimes find quoted for them, with pressures much less than a billionth of the Earth's atmospheric pressure, consist largely of stray atoms knocked of the surface by micrometeorite and cosmic ray impact. So sparse are these atoms that each is more likely to drift off into space rather than collide with another atom. This condition defines the 'exosphere' of a planet. It is merely the tenuous outermost zone of most atmospheres, but is all that the Moon and Mercury can muster.

The stronger gravity of the more massive terrestrial planets enables them to hang on to gas more effectively, although the density and chemical composition have evolved out of all recognition as a result of numerous processes. In the early days, the more active solar wind may have stripped away most of each original atmosphere, but these would be replenished by volcanism. An important ongoing process is that short-wavelength solar ultraviolet light can split molecules of water vapour into hydrogen and oxygen. Hydrogen is very light, and can escape to space, which makes this 'photodissociation' of water an irreversible process. Venus and Mars have both lost much of their original water in this way. The compositions of the present-day atmospheres of Venus, Earth, and Mars are summarized in Table 4.

Having been split by ultraviolet light, atmospheric molecules can combine with others, by series of reactions described as 'photochemistry'. This occurs especially in a planet's 'thermosphere' which begins about 100 kilometres above the surface, so named because this layer is heated by the solar ultraviolet energy used in either splitting molecules apart or stripping them of some of their electrons. The latter process is called ionization, and ions (mainly of oxygen for the Earth and carbon dioxide for Venus and Mars) can be sufficiently common in the outer reaches of a thermosphere to make an electrically conducting layer referred to as the 'ionosphere'. When a solar storm brings plasma from the Sun to the Earth, this distorts the magnetic field and causes unusual currents to flow in the ionosphere that can badly disrupt radio communications and even cause power failures.

Table 4 The present-day atmospheres of terrestrial planets, showing the abundances of the six most common gases expressed as a percentage of the total number of molecules (water is very variable in Earth's atmosphere), and the surface pressure relative to Earth

Venus		Earth		Mars	
CO_2	96.5	N_2	78.1	CO_2	95.3
N_2	3.5	O_2	20.9	N_2	2.7
SO_2	0.015	H_2O	Up to 4	Ar	1.6
H_2O	0.01	Ar	0.93	O_2	0.13
Ar	0.007	CO_2	0.034	CO	0.07
H_2	<0.0025	Ne	0.0018	H_2O	0.03
Surface pressure	92	Surface pressure	1	Surface pressure	0.0063

Deeper layers of an atmosphere, where the short-wavelength ultraviolet does not penetrate, are immune from photochemistry. Here the air is warmed mostly by contact with the ground (which is warmed directly by the Sun), and so in the lowest layer, called the troposphere, atmospheric temperature decreases upwards. Atmospheric pressure and density also decrease upwards, which means that the troposphere contains most of the mass of the atmosphere. In the troposphere, composition can evolve because of chemical reactions between air and rock (this is a corollary of chemical weathering) and especially (perhaps only) in the case of the Earth because of life. Here plants and primitive single-celled organisms use solar energy and atmospheric carbon dioxide to build their bodies, releasing gaseous oxygen that was vanishingly rare in the original atmosphere. Without plants, oxygen-breathing animals (like us) could not exist. The temperature would also be different, as I will explain shortly.

When air near the base of the troposphere is heated, it must expand, which makes it buoyant. It will then rise, to be replaced by colder air displaced from above. This is another example of convection (which you previously met within a planet's mantle), and is what drives the weather on the Earth, Venus, and Mars. The pattern of circulation is different in each case because it also depends on such factors as the planet's rate of rotation (slow for Venus), the rate of rotation of the atmosphere (much faster than that of the planet itself in the case of Venus's upper troposphere), and the day–night temperature difference (high for Mars, small for Venus). Figure 15 shows the characteristic circulation above Venus's south pole. In contrast, spiral storm systems in the Earth's atmosphere tend to begin near the tropics.

The Earth's atmosphere also differs from its neighbours in the complexity of its layering. At Venus and Mars, temperature decreases rapidly with altitude in the troposphere, then decreases more slowly with altitude in a (non-convecting) layer

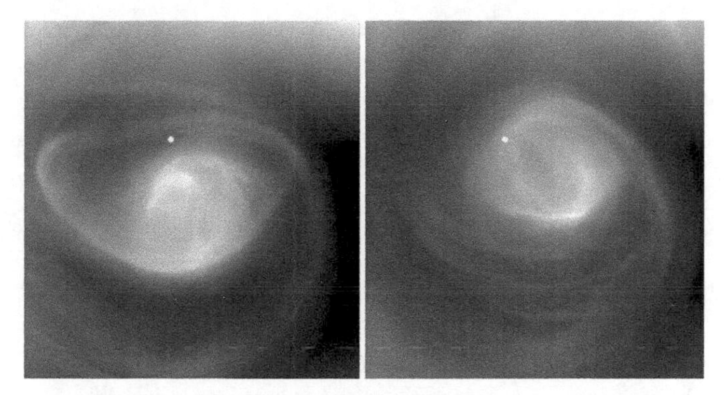

15. The 'eye' of Venus's 2,000-kilometre-diameter south polar vortex, imaged 24 hours apart. The dot indicates the south pole. These images, recorded in the mid-infrared, see the cloud-tops about 60 kilometres above the surface. The centre of the 'eye' is warmer (appearing brighter), showing that here the clouds are drawn downwards to warmer, deeper levels

called the mesosphere, and then rises with altitude in the thermosphere because of the ultraviolet absorption. The Earth is unique among terrestrial planets in having a layer extending from about 10 to 50 kilometres altitude, between its troposphere and mesosphere, where temperature increases with altitude. This is the stratosphere, which is warmed by the absorption of 230–350-nanometre-wavelength ultraviolet photons (to which the thermosphere and mesosphere are transparent) by ozone molecules. Ozone is three oxygen atoms bound in a molecule (O_3), as opposed to two oxygen atoms (O_2) which is what is usually meant when referring to 'oxygen', and is assembled from oxygen by photochemical reactions higher in the atmosphere.

Greenhouse effects and the hole in the ozone layer

Many people are aware of the 'hole in the ozone layer' and the 'greenhouse effect', but tend to conflate them as twin culprits of climate change. However, the two are very different.

The ozone layer occurs (only) in the Earth's stratosphere, and is where 230–350-nanometre ultraviolet light is absorbed. This is very important for ourselves and other surface-dwelling life, because if it is not blocked out such radiation can cause skin cancers and genetic damage. It takes surprisingly little ozone to be an effective screen. If you gathered all the ozone that is dispersed in the stratosphere into a single layer at sea-level it would be only about 3 millimetres thick. This is a fragile veil, so when in the 1970s and 1980s it became apparent that over Antarctica the stratosphere had lost perhaps half its ozone, there was considerable concern, and talk of a 'hole in the ozone layer'. The main cause was traced to reactions involving industrial chemicals called chlorofluorocarbons (CFCs) which, as a consequence, have now been banned from their former use in aerosol sprays and refrigerants so they cannot leak into the atmosphere. The Antarctic 'ozone hole' and a lesser one over the Arctic have now stabilized. Depletion of ozone is only a few per cent outside of the polar regions, and is undetectable over the tropics.

There is no simple link between ozone concentration and mean global temperature. A badly depleted ozone layer would make life unpleasant, but has little to do with climate change or global warming. The tropospheric temperature of a planet is controlled by how effectively the lower atmosphere absorbs infrared radiation. This is because visible sunlight warms the ground, and the warmed ground emits infrared radiation. The temperature of the atmosphere depends on two factors: the heat it picks up from contact with the ground, and how much of the outgoing infrared radiation it can absorb.

Most gas species are transparent to infrared radiation, but molecules consisting of two or more different elements absorb infrared quite strongly. Thus nitrogen (N_2), oxygen (O_2), and argon (Ar) do not absorb infrared, but water vapour (H_2O), carbon dioxide (CO_2), sulfur dioxide (SO_2), and methane (CH_4) do.

By analogy with trapping warmth inside a greenhouse, this is called the 'greenhouse effect'. There is a natural greenhouse effect in the atmospheres of Venus, Earth, and Mars. Thanks mostly to its enormous load of carbon dioxide, the atmospheric greenhouse effect on Venus maintains its surface temperature an impressive 500 °C above what it would otherwise be. Water vapour and carbon dioxide warm the Earth by about 30 °C, and greenhouse warming of Mars, which has a tenuous carbon dioxide-rich atmosphere, is only about 6 °C.

Earth's greenhouse effect keeps the temperature within a range to suit the life that has evolved here. Mediated by life itself, the strength of the greenhouse effect has changed to keep the temperature in the right range. Four billion years ago, the Sun was only about 70% as brilliant as it is today, so the Earth would have been very much colder, had its atmosphere been the same as today's. However, before 4 billion years ago, it was probably mostly carbon dioxide, and 100 times denser than today, so the greenhouse effect would have been stronger. Thanks to primitive algae, the carbon dioxide content had decreased to about 10 times its present value by about half a billion years ago, so of course the greenhouse effect must have declined too. Free oxygen (O_2) first appeared some time between about 2.7 and 2.2 billon years ago, and peaked at about 170% of its present concentration between 250 and 200 million years ago. Clearly, life on Earth has both influenced and benefited from changes in the composition of the atmosphere.

Since the beginning of the industrial era, human activity has affected the atmosphere in various ways: ozone depletion, industrial smog, and so on. However, what should concern us most is our release of carbon dioxide into the atmosphere, or, rather, *back* into the atmosphere, for most of this is carbon dioxide previously extracted from the atmosphere by organisms and sequestered as coal or oil. The amount of atmospheric carbon dioxide increased by about 35% in the 60 years since 1960 (faster than any natural process), and is still increasing. This 'anthropogenic greenhouse effect' will inevitably lead to warming of the global climate. A few degrees' rise in temperature will affect ecosystems and will also tend to make local weather (including short-term temperature fluctuations) more extreme. Another consequence will be a rise in global sea-level, chiefly because water expands as it warms. So, although the natural greenhouse effect in our atmosphere is a good thing, human-induced rapid increases in the effect could have potentially disastrous consequences for our civilization.

Against a background of general gradual decrease in the natural greenhouse effect, which counterbalanced the slow waxing of the Sun's luminosity, there have been several excursions in the Earth's climate. Ice Ages, when much (in extreme cases all) of the surface water was frozen, are the best-known example. These are controlled not so much by the atmosphere as by variations in the tilt of the axis and the eccentricity of the orbit. Similar effects probably explain the drastic changes in the wetness of Mars's surface over time.

Clouds

Clouds are highly reflective, so the cloudier an atmosphere, the more solar energy is reflected directly back into space. However, a cloudy sky increases the ability of an atmosphere to trap heat from the sunlight that *does* reach the ground, so the effect of clouds on global temperature is complex. The unbroken clouds of Venus have not saved its surface from being thoroughly cooked by the greenhouse effect.

Clouds form when the temperature and pressure make it favourable for some constituent of the atmosphere to condense as liquid droplets or ice particles. In the case of the terrestrial planets, the relevant constituent is usually water. Although water is only a small fraction of Venus's atmosphere, there is enough to form a continuous layer of cloud at the top of its troposphere, between about 45 and 65 kilometres above the surface. There, water vapour condenses as droplets about 2 micrometres across. These remain suspended, being too small to fall, and are described as aerosol droplets. Atmospheric sulfur dioxide dissolves in them, so they turn into sulfuric acid. However, if anyone tries to tell you that it 'rains sulfuric acid on Venus', they are wrong. Wherever the droplets are drawn down below about 45 kilometres by atmospheric circulation, the heat causes them to evaporate again, and they never have the chance to become raindrops large enough to fall as far as the ground.

Above about 6 kilometres, Earth's clouds consist mostly of tiny ice particles, and below that altitude they are mostly droplets of water. Rainclouds are not really grey, they just look that way because they are thick enough to blot out so much light. On Mars, clouds are comparatively rare. In most of its troposphere, they are made of water-ice, whereas around 80 kilometres, near the troposphere/mesosphere boundary, clouds of carbon dioxide particles have been observed.

Polar caps and oceans

As well as condensing to form clouds, atmospheric constituents may condense as either ice or liquid at the surface. The Earth is the only terrestrial planet to have oceans today, which of course are made of water. Near the poles, water is frozen to form polar caps. The young planet Venus may have enjoyed a brief epoch when oceans covered the globe, before the evaporated water vapour (subsequently lost by photodissociation) added to a burgeoning greenhouse effect leading to the current parched situation.

However, Mars is different. A vast 'Oceanus Borealis' occupying the whole of the low-lying northern plains about 3.8 billion years ago was in vogue in the 1990s. Although that remains contentious, many would accept the likelihood of lakes on Mars extensive enough to be called 'seas' at the time when channels such as those in Figure 13 were flowing, and some frozen relics may even

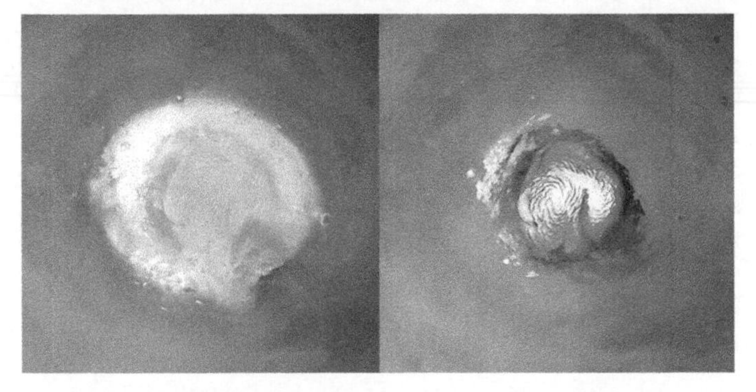

16. 1,500-kilometre-wide images of Mars's northern polar cap in early spring (left) and high summer (right). In summer, most of the carbon dioxide frost has sublimed (turned from ice to vapour), leaving only the residual, 'permanent' cap of water-ice

survive, buried by dust (Figure 10). However, there is no doubt that ice exists at the surface today in Mars's polar caps (Figure 16). These consist of 'permanent' water-ice with a fringe of carbon dioxide frost that grows and contracts seasonally.

Earth's and Mars's polar caps interact with the atmosphere. They are, in effect, deposits of gases that have 'frozen out' of the atmosphere, either falling from the clouds as snow or condensing directly onto the ground. When the temperature rises, material from the polar caps is returned to the atmosphere, either by melting and then evaporation (for water on Earth, and probably Mars in the past) or by subliming directly from ice to vapour (on Mars, for carbon dioxide and water today).

Equilibria like these cannot occur on airless bodies like the Moon and Mercury, and so polar caps are not to be expected. However, during the 1990s it was noted that radar signals are reflected with unusual strength from permanently shadowed regions inside craters near the poles of both bodies. This would be consistent with water-ice dispersed as grains within the regolith. A possible explanation is that the floors of these craters are so cold that any stray water molecules that wander into them tend to adhere to the surface in 'cold traps'. This water need not be part of these bodies' original inventory, it could have been supplied later by impacting comets. Finding a supply of water on the Moon is of great importance if a human colony, or even just a permanently occupied base, is ever to be established there. The poles are clearly the best bet, and in 2009 water was confirmed in an ejecta plume created when a spacecraft was crashed into a permanently shadowed polar crater. Infrared spectra obtained by other spacecraft revealed water and hydrated minerals dispersed in the regolith across broader regions, in minute concentrations but raising hopes that the Moon might not be so wholly inhospitable as formerly believed.

The MESSENGER mission proved conclusively that permanently-shadowed craters near Mercury's north pole have water-ice within them, rather like the Moon, but the spacecraft configuration prevented a similar proof at the south pole.

Cycles

Interplay between interior, surface, and atmosphere, and the cycling of components between them, is extremely important. The Earth's 'hydrologic cycle' is the most familiar example. It is not a single cycle, but an array of interconnected loops. In essence, water in the oceans evaporates to form clouds, and later precipitates out as rain or snow that eventually finds its way back into the oceans (via rivers or seasonal polar caps). Water can be drawn into the interior (deeply at subduction zones or more shallowly by infiltration of the ground) and re-emerge via volcanoes. It can also react chemically with rock (chemical weathering) and be stored within minerals. There is also an important 'carbon cycle' with loops passing between atmospheric carbon dioxide, living plants and animals, dissolved carbon dioxide, marine limestones, hydrocarbon deposits, volcanic gases, and so on.

Mars is sure to have similar cycles, though acting more sporadically and over different timescales, and with different relative importances for each loop. There are probably even slower cycles involving carbon dioxide and sulfur dioxide on Venus, in which the atmosphere weathers the surface rocks, which eventually become buried by lava flows to depths at which the gases are liberated once more and escape back to the atmosphere through volcanic vents. Until we have explored and documented the complexities and timescales of these multi-looped and inter-related cycles, our understanding of what makes each planet 'tick' will remain naive.

Chapter 3
Giant planets

These are the bodies that dominate the Solar System – provided you think it is size that matters, and are willing to overlook the Sun itself. The four giant planets are illustrated to scale in the lower half of Figure 3, showing how comprehensively their size overshadows the terrestrial planets. The view of Uranus is from the Hubble Space Telescope in orbit about the Earth, whereas the other giant planets are as seen by visiting spacecraft. Their domination by mass is not quite so overwhelming, because they are less dense than the terrestrial planets. Jupiter's density is only 24% of the Earth's, and Saturn is even less dense and would float if dropped into a sufficiently large (and purely hypothetical) bucket of water. All of them have rings in their equatorial plane, though only those of Saturn and Uranus are sufficiently prominent to be visible in Figure 3. Although the rings look solid, they are made of myriads of orbiting particles and are extremely insubstantial. They are discussed, along with the giant planets' satellites, in the next chapter.

By convention, the size of a giant planet is measured from the top of its clouds. These occur in the planet's troposphere, above which are largely transparent and progressively less dense layers classifiable in the same way as for the Earth's atmosphere. The base of a giant planet's troposphere is hard to define and has never been explored even in the case of Jupiter, where in 1995 an entry probe released by the *Galileo* spacecraft reached a depth of 160 kilometres below the cloud-tops before pressure (22 atmospheres) and temperature (153 °C) put paid to it. Probably, the troposphere of each giant planet merges seamlessly into a fluid interior at temperatures and pressures so high that there is no distinction between gas and liquid. Certainly, there is no solid surface that a human could ever stand upon.

Basic data for the giant planets are given in Table 5. The polar diameters quoted there are less than equatorial diameters, because rapid rate of rotation (see Table 2) flattens their shapes. Jupiter's polar diameter is 6.5% less, and Saturn's 10% less, than its equatorial diameter. The difference is only about 2% for the less gassy and more slowly rotating Uranus and Neptune (and is less than 1% for each terrestrial planet).

Interiors

There is no simple way to study the interior of a giant planet, but we can use atmospheric composition (99% hydrogen and helium) and our general knowledge of what the Solar System as a whole is made of to construct a model that is consistent with its measured density, and with the interior pressures that we can infer from this. Below the atmosphere, each giant planet must have a zone consisting mostly of hydrogen molecules (H_2) and helium atoms (He), in a state that it is better to call 'fluid' rather than either 'liquid' or 'gaseous'. At the very centre, there is probably a rocky inner core, of about ten Earth-masses inside Jupiter, three inside Saturn, and one inside Uranus and Neptune. Surrounding the inner core, there ought to be an outer core of 'ice' composed of unknown proportions of water, ammonia, and methane, amounting to about five Earth-masses inside Jupiter, maybe six Earth-masses inside Saturn, twelve inside Uranus, and fifteen inside Neptune. We do not know whether these outer and inner cores are molten or solid, because although we can estimate the pressure (a staggering 50 million

Table 5 Basic data for the giant planets. Note that the mass units are a thousand times bigger than for the terrestrial planets in Table 3

	Mass/ 10^{27} kg	Polar diameter / km	Density / 10^3 kg m^{-3}	Cloud-top gravity / m s^{-2}	Cloud-top temperature
Jupiter	1.90	133,700	1.33	23.1	−150 °C
Saturn	0.569	108,720	0.69	9.0	−180 °C
Uranus	0.0868	49,940	1.32	8.7	−214 °C
Neptune	0.102	48,680	1.64	11.1	−214 °C

atmospheres in the centre of Jupiter), we do not know the composition and have only a vague idea of the likely temperature (ranging from in excess of 15,000 °C in the centre of Jupiter to about 2,200 °C at the outer edge of Neptune's core). Our understanding of how materials behave under such extreme conditions is sketchy, including whether metallic iron could differentiate from the rock and sink towards the centre to form an inner-inner core. The cores of Uranus and Neptune might even be undifferentiated mixtures of ice and rock.

Accounting for the cores leaves little more than one Earth-mass for the hydrogen and helium exteriors of Uranus and Neptune, comprising shells about 6,000 kilometres thick. However, the 'gas giants' Jupiter and Saturn have much deeper envelopes of hydrogen and helium surrounding their cores, in excess of 300 and 80 Earth-masses, respectively. Hydrogen is easier to model than ice or rock, and scientists are pretty confident that at pressures greater than about 2 million atmospheres, hydrogen atoms are squeezed so tightly together that electrons are no longer confined about specific atoms. Instead, they are able to wander through a sea of hydrogen that behaves like a molten metal. This freedom of electron movement makes 'metallic hydrogen' an excellent conductor of electricity. A shell of metallic hydrogen (with some helium dissolved in it) surrounding Jupiter's core probably accounts for about 250 Earth-masses (80% of Jupiter's total mass), whereas around Saturn's core it is thought to comprise a more modest 41 Earth-masses (just over 40% of Saturn's total mass). Figure 17 illustrates the full internal structure of Jupiter.

The internal structure of the giant planets may still be evolving because, with the possible exception of Uranus, they all radiate more heat to space than they receive from the Sun. Jupiter is so massive that it could still be leaking out a significant amount of primordial heat trapped within since its formation, but for Saturn and Neptune this heat excess shows that heat must actually be

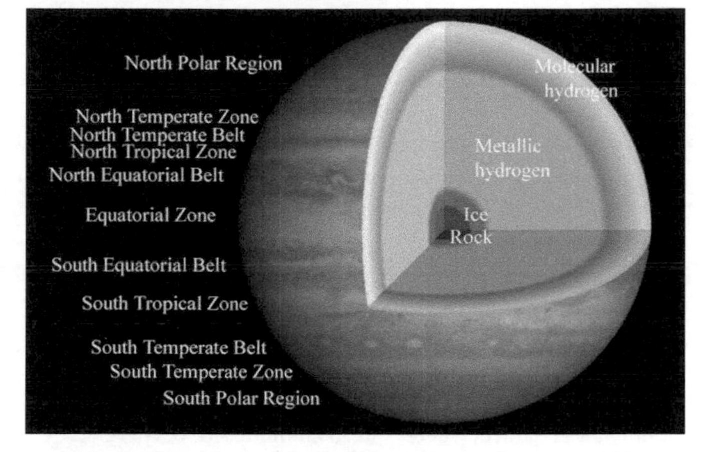

17. Cut-away diagram showing the proposed internal layers within Jupiter. The principal tropospheric cloud-top zones (bright) and belts (dark) are labelled

being generated within. The discrepancy is too large to be radiogenic heat, so internal differentiation may still be occurring. The settling of denser than average material inwards (allowing an inner shell to grow while the surrounding shell becomes thinner but purer) would convert gravitational potential energy into heat. Such heat could come from continuing growth of cores (or inner cores) or, for Saturn only, from inward settling of helium droplets inside its metallic hydrogen layer.

Atmospheres

Composition

In contrast to the reasoned speculation about giant planet interiors, understanding of their atmospheres can draw more on observation and measurement. The composition of clouds and the overlying layers can be measured by optical spectroscopy, which is the study of how sunlight of different wavelengths is absorbed at various depths within the atmosphere. In addition, the average molecular mass at each depth can be determined by the amount of refraction experienced by radio signals transmitted by a spacecraft while it disappears from view behind the planet. Also, the *Galileo* entry probe made various measurements inside Jupiter's atmosphere during its descent. Table 6 compares the chemical composition of the four giant planets' atmospheres. In addition to the species listed there, each contains smaller traces of ethyne (C_2H_2), Jupiter has ethene (C_2H_4), and Jupiter and Saturn both have phosphine (PH_3), carbon monoxide (CO), and germane (GeH_4).

The topmost layer of continuous clouds on Uranus and Neptune is of methane-ice particles. It is too warm for methane condensation at Jupiter and Saturn, where instead ammonia-ice particles condense to form the topmost clouds. These top cloud layers are about 10 kilometres thick, below which the 'air' probably becomes clear again. Calculations suggest that in the case of Jupiter, there should be a second layer of clouds made of ammonium hydrosulfide (NH_4HS) about 30 kilometres below, and a third cloud layer, this time of water (ice at the top, liquid water droplets below) about 20 kilometres lower still. The *Galileo* entry probe found probable ammonium hydrosulfide clouds at about the right depth but did not find any water-ice clouds. Some say the models

Table 6 Gases detected in the atmospheres of the giant planets, showing the measured proportion made up by each

	Jupiter	Saturn	Uranus	Neptune
Hydrogen (H_2)	0.90	0.96	0.83	0.80
Helium (He)	0.10	0.03	0.15	0.19
Methane (CH_4)	3×10^{-3}	4.5×10^{-3}	0.023	0.015
Ammonia (NH_3)	3×10^{-4}	1×10^{-4}	-	-
Water (H_2O)	4×10^{-6}	$<2 \times 10^{-9}$	-	-
Hydrogen sulfide (H_2S)	$<1 \times 10^{-7}$	$<2 \times 10^{-7}$	-	-
Ethane (C_2H_6)	6×10^{-6}	2×10^{-7}	-	1.5×10^{-6}
Visible clouds	Ammonia	Ammonia	Methane	Methane

are wrong; others say that the probe penetrated into a gap between discontinuous water-ice clouds. The same cloud layers are expected at Saturn, but spaced about three times further apart because of Saturn's lower gravity. Ammonia-bearing clouds are expected below the methane clouds of Uranus and Neptune.

The atmospheric pressure at the top of Jupiter's ammonia clouds is a factor of two or three less than the sea-level atmospheric pressure on Earth, whereas on the other giant planets the cloud-top pressure is close to Earth's sea-level pressure.

Circulation

A global pattern of cloud bands running parallel to the equator is visible on Jupiter even through a small telescope. A similar pattern is repeated less dramatically on the other giant planets. Solar heating must play some role in the circulation of this, visible, part of their atmospheres, but it appears to be powered mostly by internal heat and to be controlled by their rapid rotation.

Traditionally, the dark bands are referred to as 'belts' and the intervening bright bands as 'zones'. The names given to the main belts and zones on Jupiter are indicated on Figure 17. Because there is no solid surface to act as a frame of reference, wind speeds on giant planets are measured relative to the planet's average rate of rotation. On Jupiter, the cloud-top wind blows to the east at up to 130 metres per second across most of the Equatorial Zone. The adjacent edges of the North and South Equatorial Belts share this motion, but the wind speed decreases and ultimately reverses with distance away from the equator across each belt until the Tropical Zones are reached, where the wind direction reverses again, and so on with repeated reversals across each belt and zone until the polar regions.

In Jupiter's zones, the atmosphere is mostly rising, leading to condensation of ammonia clouds high up, where they naturally appear bright. Conversely, in the belts the atmosphere is mostly sinking, drawing the cloud-tops lower, so to a depth where they look darker. Local exceptions to this pattern have been identified on Jupiter, and the general rule of rising zones and sinking belts scarcely seems to hold at all on the other giant planets, whose atmospheric circulation is harder to fathom. A complicating factor influencing the visibility of zones and belts is the poorly understood nature and abundance of whatever trace compounds add colour to the clouds, and which are expected to result from photochemical reactions. Jupiter's various hues of yellow and red could be caused by sulfur (released photochemically from either hydrogen sulfide or ammonia hydrosulfide), phosphorous (from phosphine), or hydrazine (N_2H_4, made photochemically from ammonia).

Colour variations are less pronounced in Saturn's atmosphere, and the pattern of zones and belts is less prominent. However, wind speeds are higher, with eastward-blowing winds in excess of 400 metres per second prevailing for 10° either side of the equator.

Rotating storm systems are well known on both Jupiter and Saturn. The most famous is Jupiter's Great Red Spot. This can be seen in Figure 3, as an oval feature straddling the boundary between the South Equatorial Belt and the South Tropical Zone. It extends 26,000 kilometres from east to west, having a spiral structure and taking about six days to rotate anticlockwise. It has been apparent in telescopic observations since at least 1830. Smaller storms can be made out at a variety of scales on both Jupiter (look along the South Temperate Belt in Figure 17) and Saturn. About once every 30 years, during summer in its northern hemisphere, Saturn tends to be disfigured by a giant storm system that begins as a white spot near the equator, but within a month can spread to encircle the globe before gradually fading from view.

Whereas Jupiter and Saturn have a yellowish cast, Uranus and Neptune appear bluey-green. This is because we see their cloud-tops through a depth of overlying methane gas, which preferentially absorbs the longer (red) wavelengths of light.

The 82.1 °C axial tilt of Uranus makes for extreme seasonal variations. For example, when *Voyager 2*, the only spacecraft yet to have visited Uranus, flew past in 1986, the south pole was in full sunlight and most of the northern hemisphere was suffering decades of darkness. Its southern hemisphere looked disappointingly bland on the *Voyager* images, but as the Uranian year progressed and the Sun began to rise and set over a wider range of latitudes, the globe began more to resemble the other giant planets (Figure 18). In 2007, Uranus passed through its equinox, and the south pole, followed gradually by the rest of the southern hemisphere, began to drift into long-term darkness, which will peak with southern midwinter in 2028.

When Neptune was revealed in detail during *Voyager 2*'s 1989 fly-by, it resembled a blue version of Jupiter. There was even a giant storm system in the form of a dark spot just south of the equator, which was dubbed the 'Great Dark Spot' in tribute to its famous Jovian cousin. However, it proved to be shorter-lived, and had vanished by 1994. Unlike at Jupiter and Saturn, the equatorial wind stream on Neptune blows west (opposite to the planet's rotation), as can be seen by the westward drift of the Great Dark Spot relative to the smaller, more southerly spot in Figure 18.

Magnetospheres

Each of the giant planets has a strong magnetic field. The 'magnetic dipole moment' of Neptune, which is the conventional measure of a planetary magnetic field, is 25 times greater than the Earth's. Uranus's is 38 times, Saturn's is 582 times, and Jupiter's is 1,949 times greater. To generate these fields, each planet must contain a zone of electrically conducting fluid undergoing some kind of convective motion. In the two terrestrial planets with magnetic fields (Mercury and Earth), the explanation is a fluid

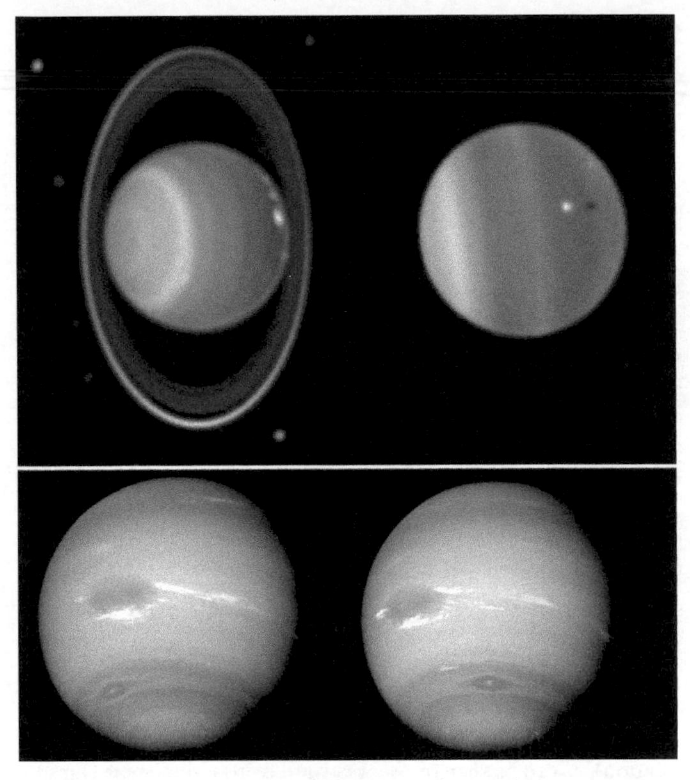

18. Top: Uranus seen by the Hubble Space Telescope in August 1998 (left) and July 2006 (right). The change in orientation of the planet's axis relative to the Sun is apparent from the pattern of the atmospheric banding. The region around the south pole was still in sunlight in 1998, but the axis had become nearly edge-on to the Sun by 2006. High, bright clouds are apparent in the far north in the 1998 image, which also shows the rings and several of the inner satellites. The rings were edge-on and invisible in 2006, but instead we can see one of the regular satellites (Ariel) and its shadow. Bottom: two images of Neptune seen by *Voyager 2* during its approach in 1989, recorded almost exactly one planetary rotation apart. The Great Dark Spot and associated wisps of high, bright nitrogen cirrus clouds are prominent. Note also the general banded structure, and a smaller dark spot further south

shell of their iron cores. The magnetic fields of Jupiter and Saturn are probably generated in the metallic hydrogen layer, stirred into motion by the planets' relatively rapid rotation. Pressures are too low for metallic hydrogen in Uranus and Neptune, so their magnetic fields are harder to account for, but are probably caused by motion within electrically conducting 'ice' of their outer cores.

An important consequence of a planet having a magnetic field (which applies to Mercury and the Earth too) is that it cocoons the planet inside a zone into which magnetic field lines from the Sun cannot usually penetrate. This zone is called the planet's 'magnetosphere'. The paths of charged particles in the solar wind (chiefly protons and electrons) are controlled by the Sun's magnetic field, until they hit the 'bow shock' of a planet's magnetosphere, which diverts them past the planet.

Charged particles can get through sometimes, especially by leaking back up the long magnetotail, down-Sun from the planet. Near the poles, these can be channelled along field lines towards the top of the atmosphere, where their arrival causes glows in the sky called aurorae, well known on Earth and observed also on Jupiter and Saturn.

Chapter 4
Giant planets' satellites and rings

Rings and a large family of satellites are features common to all four giant planets. There are variations in emphasis and scale, but the similarities between each ring-satellite system outweigh their differences.

Ring-satellite systems

Most outer satellites of each giant planet travel in eccentric orbits, usually in the opposite direction to the spin on their planet. Furthermore, many of these orbits are inclined at >30° relative to their planet's equator. The typically eccentric, retrograde, and inclined nature of their orbits earns these bodies the name 'irregular satellites', quite apart from the fact that (being at most about 100 kilometres, and more often only a few kilometres, across) they have far too little gravity to pull themselves into spherical shapes. The irregular satellites are the most numerous class: at the last count, Jupiter had 71, with orbital semi-major axes ranging from 105 to 400 Jupiter radii; Saturn had 38, with orbits from 184 to 417 Saturn radii; Uranus had 9, 167–818 Uranus radii; and Neptune had 6, 223–1,954 Neptune radii.

The 'regular satellites' are the large ones in near-circular prograde orbits, much closer to their planets, and with very low inclinations. Jupiter has 4 (the ones discovered by Galileo) whose orbital semi-major axes range from 5.9 to 26.3 Jupiter radii. These are substantial worlds, and geologically have much in common with the terrestrial planets, though of course they do not satisfy the IAU definition of a planet. Saturn has 8 (all but one considerably smaller than Jupiter's, and orbiting at 3–59 Saturn radii), and Uranus has 5 (at 5–23 Uranus radii). Neptune has one large satellite, Triton, orbiting at 14 Neptune radii, that would be regarded as 'regular', except for its retrograde orbit. An important characteristic, shared by all regular satellites (including Triton), is that tidal forces have such a grip on them that they are in synchronous rotation, rotating once per orbit, so that (like the Earth's Moon) they keep the same face towards their planet.

Closer still, we find irregular-shaped lumps of debris that it is convenient to distinguish as 'inner moonlets'. These have circular, prograde, equatorial orbits. So do the particles that make up the rings, and, given that some inner moonlets' orbits lie within the rings, there is probably no fundamental difference between a large ring particle and a small inner moonlet. Jupiter has only 4 known inner moonlets, but Saturn has 14, counting 7 whose orbits lie among those of its innermost regular satellites. Uranus has 13 and Neptune 6.

The width and number of rings varies from planet to planet, Saturn's being by far the most spectacular, but in general their thickness is no more than a few tens of kilometres. Mostly, they are closer to their planet than a distance known as the 'Roche limit', a boundary within which any large body should be ripped apart by tidal forces. Most rings are regarded as debris left over from the tidal disruption of a satellite or comet that strayed too close to the planet, but some less substantial rings are demonstrably supplied from nearby satellites by particles vented actively into space or thrown up by impacts.

Saturn's rings are made of ice and reflect about 80% of the sunlight falling on them. Despite their prominent appearance (Figure 3), the material in them would suffice only to make a body about 100 kilometres in diameter if it could all be gathered together. Although individual ring particles have not been imaged directly, the rate at which the rings cool when the shadow of their planet falls across them shows that Saturn's rings are mostly particles between about a centimetre and 5 metres in size. In contrast, Jupiter's much less substantial rings are made largely of micrometre-sized particles that are also much less reflective than

19. A 5,000-kilometre-wide view of part of Saturn's ring system, seen by *Cassini* on 27 July 2009. At this scale, the curvature of the rings round the planet (out of view to the right) is scarcely discernible. The rings reflect most sunlight where particles are most densely packed, and black space shows through in particle-free gaps. Pan, a 28-kilometre-diameter shepherd satellite, can be seen orbiting in the widest gap. As well as sweeping most of this gap clear, Pan influences narrow and discontinuous rings within the gap. The exceptional length of Pan's shadow on the rings to its right is because this image was recorded when the Sun lay very close to the plane of the rings

the bright icy lumps in Saturn's rings. The ring-material at Uranus and Neptune reflects sunlight poorly (like Jupiter's ring-material) but is mostly centimetres to metres in size (like Saturn's ring-material).

Orbital resonances lead to a complex gravitational interplay between rings and the inner moonlets that orbit among them (Figure 19). Those are often called 'shepherd satellites' because some sweep clear many of the gaps in the rings, and others form, deform, and maintain narrow rings with orbits just within or just beyond their own.

In general, rings occur closer to their planet than the regular satellites, but Saturn is an exception in that it also has a diffuse outer ring of dark, dusty material centred around the orbit of Phoebe, one of its innermost irregular satellites. The material in this ring, which was discovered in 2009 using a space-based infrared telescope, is presumably being supplied from Phoebe in some way yet to be understood.

Remarkable satellites

There was a time when pretty much everyone expected even the largest of the outer planets' satellites to be dreary objects. Ancient ice-balls, heavily pock-marked by impacts, they would record the outer Solar System's bombardment history, but would be of no further interest unless you wanted to study mutual orbital evolution. That was the standard view until 2 March 1979, when Stanton Peale, working at the University of California, published (with two colleagues) a paper pointing out that the exact 2:1 orbital resonance between Jupiter's innermost Galilean satellites, Io and Europa, ought to result in so much tidal distortion of Io's shape that its interior should be molten. From estimates of density plus spectroscopic analysis of their surfaces, it was already known that Io has a rocky crust, unlike the other satellites that are predominantly ice. To suggest a molten interior inside a rocky body (where the melting temperature is so much higher) was a particularly bold step. Few might have believed this claim if *Voyager 1* had not flown past a few days later and transmitted pictures of explosively erupting volcanoes, topped by 300-kilometre-high eruption plumes.

Although tidal heating of Io is by far the strongest, the same process affects various other satellites, and many more bear signs of ancient episodes of tidal heating. This makes them varied, and intriguing to geologists. They don't mind that in most of them only the core is made of rock, surrounded by a thick mantle of ice, with perhaps a chemically distinct icy crust at the surface. Under the low surface temperatures prevailing in the outer Solar System (ranging from –140 °C for Jupiter's satellites down to –235 °C for Neptune's satellites), the mechanical properties and melting behaviour of the ice are very closely analogous to how rock behaves in the inner Solar System. In other words, those bodies have both the behaviour and structure of a terrestrial planet, with rock in place of iron in the core, and ice instead of rock in the crust and mantle.

Io is an exception in being ice-free with a rocky crust and mantle surrounding an iron core, and would be classifiable as a terrestrial planet if it was orbiting the Sun instead of Jupiter. Europa is a hybrid, having a structure like Io buried below 100–150 kilometres of ice. Here I describe both of those and some of the other satellites that fascinate me most, concentrating on the more active examples, though even the crater-pocked ice-balls have turned out to be more interesting than the dull globes formerly imagined.

Io

Io is only slightly bigger (3,642 kilometres in diameter) and denser than our own Moon, but the two could hardly be more different. Io's terrain is resurfaced so rapidly by volcanic processes that not a single impact crater is to be seen, despite the fact that the effect of Jupiter's gravity in focusing stray projectiles inwards must mean that Io is struck more often than Jupiter's heavily cratered satellites Ganymede and Callisto that orbit beyond Europa. When the first *Voyager 1* colour close-ups of Io were studied in 1979, its yellow hue led many to suppose that the lobate lava flows that could be recognized on its surface were made of sulfur. However, it is now accepted that Io's volcanism is molten silicate material (true 'rock'). The temperatures recorded in the heart of erupting volcanic vents are well in excess of 1,000 °C, despite the intense cold beyond the active areas. The gas that escapes to drive explosive eruptions such as those in Figure 20 is mostly sulfur dioxide (whereas on Earth, it would be mostly water vapour), and both sulfur and sulfur dioxide condense on the surface as 'frost' that imparts the colour to Io.

Io lies within a belt of charged particles confined by Jupiter's magnetic field. The radiation there is so intense that NASA's *Galileo* mission controllers did not allow the spacecraft to make repeat close fly-bys of Io, so only a small fraction of Io's globe was imaged well enough to show details below a few hundred metres in size. On the most detailed images, the pixels are only 10 metres across, and even on those no impact craters have been found.

If Io's present-day rate of volcanism is representative of the long term, then its entire crust and mantle must have been recycled many times over. Covering of older surfaces by lava flows and fall-out from eruption plumes, amounting to a globally averaged burial rate of a couple of centimetres per year, obscures impact craters too rapidly for any to be apparent. If Io ever had an outer layer of ice, volcanic activity has long since vaporized it, allowing it to be lost to space, because Io's gravity is too weak to hold on to water vapour or other light gas. What a fantastic place for a volcanologist to visit, if only the harsh radiation environment did not make Io's surface so thoroughly inimical to human exploration.

20. Top: part of the crescent view of Io seen by the Pluto-bound *New Horizons* mission passing Jupiter in March 2007. The plume from a volcanic vent at a site called Tvashtar caldera on the night-side rises 300 kilometres so that its upper part is in sunlight. An incandescent glow can be seen at its source, and the shadowed lower part of the plume is faintly illuminated by light reflected from Jupiter. Bottom: A 250-kilometre-wide view of Tvashtar seen eight years earlier by the *Galileo* orbiter. Sunlight is from the left. The darkest material is recent lava flows, and the east–west bright streak near the upper left is incandescent lava being erupted from a volcanic fissure

Europa

Europa (3,130 kilometres in diameter) is my favourite. *Voyager* images from fly-bys in 1980 and 1981 showed its surface looking like a cracked eggshell, with very few impact craters to be seen. Clearly, tidal heating was somehow refashioning Europa's icy outer layer, though not so rapidly as Io. Higher-resolution imaging by the *Galileo* mission revealed a complex surface history, and led to an unusually bitter controversy. It was already well known that Europa's surface is predominantly water-ice, and the globe's overall density shows that its icy carapace has to be about 100–150 kilometres thick, overlying a denser, rocky interior. However, density arguments cannot distinguish between solid ice and liquid water. The surface ice is strong and brittle, thanks to its low temperature. The controversy that emerged was over the state of the 'ice' below the surface. Was it frozen all the way down to the rock or was the lower part liquid, capped by a floating ice shell?

The latter requires a greater rate of internal tidal heating coupled with the exotic concept of a global ocean of liquid water below the ice. So far as I am concerned, evidence from images such as Figure 21 makes it clear that the ice is generally thin, only a few kilometres, and so must be floating on water. However, for several years of *Galileo*'s orbital tour of the Jupiter system, a powerful lobby group on the imaging team persisted in trying to explain the surface features as a result of processes driven by solid-state convection in the thick ice layer.

What is now the generally accepted basis of Europa's geology is best explained by reference to Figure 21. This shows numerous high-standing 'rafts' of ice, bounded by 100-metre cliffs. The surfaces of the rafts are characterized by a pattern of ridges and grooves, running in a variety of directions. Between the rafts, the texture is more jumbled, and lacks a clear pattern. There are large expanses of Europa (beyond this region) that have not been broken into rafts and where the surface pattern is uninterrupted ridges and grooves. The rafts in Figure 21 are clearly broken fragments of this sort of terrain. The ridge and groove pattern is caused by the opening and closing of cracks, probably on a tidal cycle coincident with Europa's 3.6-day orbital period. Globally, only a few cracks would be active at any one time. When an active crack is opened (to a width of perhaps just a metre or so), water is drawn up from below. The water temporarily exposed to the cold vacuum of space at the top of the crack simultaneously boils and freezes, but pretty soon becomes covered by slush. When the crack closes, some slush is squeezed out onto the surface, forming a ridge above the closed crack. The next time the crack opens, the ridge is split, and is added to by more slush when the crack closes again. A few years of opening and closing would suffice to surround central grooves by ridges of the size we see. Eventually,

21. A 42-kilometre-wide close-up of part of the Conamara Chaos region of Europa, where 'melt-through' from the underlying ocean has allowed rafts of ice to drift apart before the area refroze. Sunlight is from the right

each crack seals permanently, but a new crack will begin to operate somewhere else, and so the pattern is built up, giving the ridge and groove terrain covering much of Europa, an appearance that has been likened to that of a ball of string.

In Figure 21, 'ball of string' terrain has been disrupted by the other great process that affects Europa. This is 'melt-through', and results in a jumbled mixture of broken ice rafts described as 'chaos'. Under a future chaos region, the ocean becomes unusually warm – maybe because of silicate volcanic eruptions on the ocean floor – and the base of the surface shell of ice gradually melts, so the ice becomes thinner. Eventually, melting reaches the surface, and rafts (or floes) of ice break away from the exposed edges of ice shelf and drift into the exposed ocean. Any exposed water would refreeze pretty quickly, and perhaps it is better to think of kilometre-thick ice rafts nudging their way into a sea covered by icy slush, rather than into truly open water like the summer thaw of pack ice in the Earth's Arctic ocean. In the north-west part of Figure 21, you can see the way many of the rafts originally fitted together, because they have not drifted far apart and their 'ball of string' textures can be matched.

After the temporary heat excess dies away, the ocean refreezes and the rafts cease to drift. The ice of the refrozen sea surface and beneath the rafts begins to thicken again. When the refrozen area is sufficiently thick and brittle, new cracks may open, and a new generation of 'ball of string' texture begins to overprint the whole region. In Figure 21, there is a young crack, flanked by a narrow ridge on either side, running diagonally across. It looks unremarkable where it crosses rafts, but you can tell that it must be young because it cuts the refrozen sea lying between the rafts.

If this story is even remotely correct, then there are some very thought-provoking implications. Chemical reactions with the underlying rock would make the ocean salty – though the most abundant dissolved salt might be magnesium sulfate rather than sodium chloride as in Earth's oceans. Any such ocean overlying tidally heated rock provides a habitat for life equivalent to where life is believed to have begun on Earth. Lack of sunlight is no hindrance, because the 'primary producers' at the base of the local food chain would derive their energy from the chemical imbalance supplied to the ocean at submarine hot springs (hydrothermal vents). Such life is described as chemosynthetic, as opposed to photosynthetic. On Earth's ocean floors, the hottest vents are called 'black smokers' because of the plume of metal sulfide particles that forms when the vent fluid mixes into the ocean water. These vents are surprising oases of life, where communities of organisms (including some as advanced as shrimps and crabs) feed on chemosynthetic microbes that gain their energy by converting carbon dioxide into methane. If life on Earth began in such a setting, why not also on Europa?

Life sealed below ice that is normally kilometres thick would be extremely challenging to find, requiring Europa landers to drill or melt a hole through the ice in order to launch a robotic submarine probe that could home in on a 'black smoker' plume. However, such an ambitious mission may not be necessary if the ridges either side of a young crack are built of slush squeezed up from the ocean. While a crack is open, it could provide a niche for photosynthetic life such as plants or (more reasonably) marine algae. Like life on Earth, these could have evolved from chemosynthetic forebears. Radiation would render the top few centimetres of the exposed water column uninhabitable, but there would be enough sunlight for photosynthesis in the next few metres. If there are primary producers (plants, algae) living off sunlight, there could be animals feeding on them. To find out, the first step is to investigate a sample from the ridge squeezed out of a crack.

Both NASA and ESA have missions planned that should arrive to study Europa (and other moons of Jupiter) in about 2030. These are Europa Clipper and JUICE. They will map the ice thickness, search for biogenic molecules among the slush squeezed out of cracks, and sample any material vented from plumes emanating from cracks in the ice, which were first reported using the Hubble Space Telescope in 2016. Both missions will orbit Jupiter; any Europa lander would be at least a decade later.

Enceladus

It would be much easier to find biomarkers if Europan ice could be sampled without having to go down to the surface. Enceladus, a satellite of Saturn, would be a more reliable target. It is only 504 kilometres in diameter and has too low a density to contain much rock. *Voyager* showed it to be a strange little world, heavily cratered in parts but elsewhere apparently lacking in craters. The higher-resolution images transmitted by *Cassini*, which began an orbital survey of the Saturn system in 2004, shows a surface cut by many families of cracks (though rather unlike the 'ball of string' regions of Europa). It also discovered jets of icy crystals venting to space from cracks near the south pole (Figure 22). Fortunately, *Cassini* carried a mass spectrometer intended for study of ions and neutral particles, so the spacecraft's trajectory was adjusted to

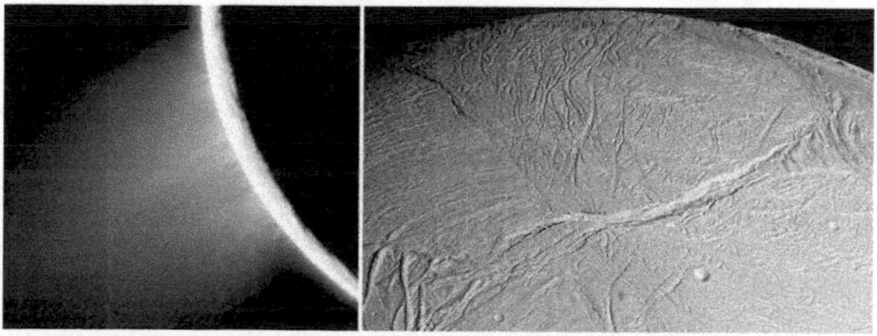

22. Two *Cassini* images of Enceladus. Left: an overexposed crescent view, showing a plume extending at least 100 kilometres above the surface. Right: an oblique view across part of Enceladus cut by several families of cracks, like those from which the plume is known to originate. A few small impact craters (too small for *Voyager* to see) show that this particular region is probably no longer active

allow it to fly through the plume and capture some samples. These were found to contain water, methane, ammonia, carbon monoxide, carbon dioxide and simple organic molecules. It also found tiny particles of silica and molecular hydrogen, which are strong indicators that water has reacted with hot rock in a setting that would be very favourable to some forms of microbes known on Earth.

Almost certainly, tidal heating (driven by 2:1 orbital resonance with Saturn's next-but-one satellite Dione) drives the crack formation and provides the impetus for the plumes. However, no one expected Enceladus to be so active, and this is particularly baffling given that its similar-sized neighbour Mimas is an archetypical cratered ice-ball showing no history of activity. It seems that Enceladus has an internal global ocean just like Europa, so it is well worth exploring further. Liquid water is good for life, but the availability of nutrients within Enceladus is surely much more restricted than within a large body such as Europa, so Enceladus does not seem such a promising habitat.

Titan

Titan is Saturn's only satellite that rivals Jupiter's Galilean satellites in scale (5,150 kilometres in diameter). *Voyager* showed it only as a fuzzy orange ball, because – alone among satellites – it has a dense atmosphere. This is 97% nitrogen but is made opaque by methane and its photochemical derivatives that turn the stratosphere into an opaque smog. Titan has a crust and mantle made of ice (mostly water-ice) occupying the outer one-third of Titan's radius and overlying a rocky core. There could be an iron inner core, in which case, to balance out the average global density, the base of the icy mantle would have to be deeper. Titan's rotation period is affected by seasonal winds, showing us that the lithosphere must be decoupled from the interior, most likely by an internal ocean. This could be mostly water or a mixture of water and ammonia (which can remain liquid at a considerably lower temperature than pure water). Most models place it as a layer *within* the icy mantle, rather than situated immediately on top of the internal rock.

The *Cassini* mission tackled the problem of seeing through to Titan's surface in three ways: it obtained blurred but usable images of the surface in some narrow bands of the near-infrared spectrum where the smog is least opaque, it used imaging radar like the *Magellan* Venus orbiter to see the ground irrespective of clouds, and it carried a landing craft, named *Huygens*, that provided images from below the clouds during parachute descent to the surface. Titan's surface geological processes revealed by this array of imaging techniques are superb analogues to many of the processes that occur on Earth. The crust is dominantly water-ice, very rigid and rock-like in its behaviour in Titan's −180 °C surface environment. *Huygens* came to rest near the equator on a sandy plain strewn with pebbles. It looked like Mars except that both sand and pebbles were made of ice. The sand could have been wind-blown, and indeed radar images reveal vast fields of wind-blown sand dunes in other parts of Titan. However, the pebbles must have been transported by flowing liquid, which, given Titan's atmospheric composition and surface temperature, must be methane (CH_4) or ethane (C_2H_6). As it descended, *Huygens* saw branching drainage channels near to the landing site, and radar imaging reveals complex valley systems in many other regions, starting in highlands where the 'bedrock' of icy crust is exposed and draining into lowland basins where sediment accumulates. Better than that, it found lakes of ethane-tainted liquid methane near both poles (Figure 23). Some lake-beds were dry, and others had shallow or marshy fringes, and it is likely that they vary seasonally. Titan is clearly geologically active. A few deeply eroded impact craters have been recognized, and there are some sites of suspected 'cryovolcanism', where icy 'magma' is erupted analogously to terrestrial lava flows.

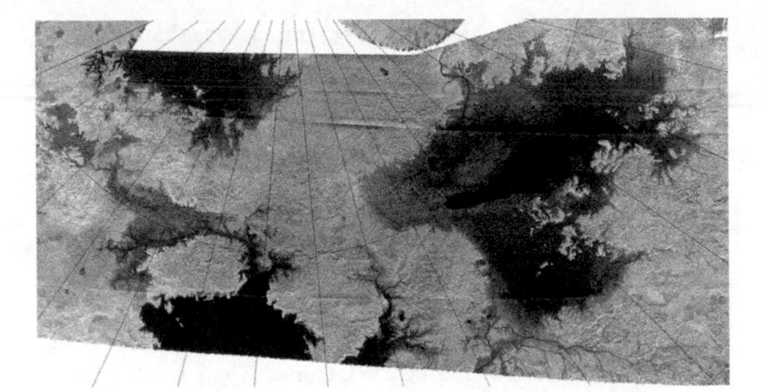

23. A 1,100-kilometre-long mosaic of *Cassini* radar images, near Titan's north pole. The dark areas are lakes, the largest of which exceeds 100,000 square kilometres, 20% bigger than Lake Superior in North America. Dendritic drainage channels can be seen feeding the lakes. Lines of longitude have been added; blank areas are unimaged

The extent to which cryovolcanism and tectonic processes contribute to the sculpting and resurfacing of Titan's surface is unknown. However, it is clear that erosion of bedrock (in this case, ice, of course) followed by transport and deposition of sediment are major players. Rainfall on Titan must consist of droplets of methane that, like rainfall on Earth, infiltrates the ground and feeds springs that supply streams and rivers. The capacity of methane to react chemically with the icy 'bedrock', its erosive power, the rate at which it evaporates back into the atmosphere, and how long it remains there before raining out again are uncertain. All these must be factors in a methanologic cycle that mimics Earth's hydrologic cycle. Mars had rainfall, rivers, and lakes long ago, but Titan is the only other place where they occur today. One day, we will send another probe to explore Titan more thoroughly – perhaps including a balloon to drift below the smog, with variable buoyancy so it can touch down in interesting places. Such a mission could sample the lake fluid, and obtain pictures of waves breaking on a thoroughly alien shore.

Miranda and Ariel

Although present-day cryovolcanism on Titan remains controversial, ancient cryovolcanism cannot be doubted on two of Uranus's five regular satellites, Ariel and Miranda, where the surface temperature is –200 °C. Its effects can be seen on images sent back by *Voyager 2*, which flew through the Uranus system in January 1986.

Ariel is the larger of the two (1,158 kilometres in diameter). It is a complex globe, whose oldest cratered terrain is cut by numerous faults bounding high-standing blocks. Most of the faults define flat-floored valleys of the kind denoted by the descriptor term 'chasma'. However, rather than preserving down-dropped highland surface, the floors of most chasmata have been covered by smooth material, or at least by something that appears smooth at the 1-kilometre resolution of the *Voyager* images.

Probably in the distant past (more than 2 billion years ago), tidal heating led to fracturing of Ariel's surface and the effusion of cryovolcanic lava. This covered the floors of the chasmata, and in places can be seen spreading beyond them to partly bury some older impact craters. At the distance of Uranus from the Sun, its ice is expected to be a more complex cocktail than the slightly salty ice found on Jupiter's satellites. The most likely melt to be extracted by partial melting is a 2:1 mixture of water and ammonia. This is liquid at –100 °C and so can be generated by much more modest heating than would be required to melt pure water.

Individual 'lava' flows can also be seen on Miranda, which is Uranus's smallest regular satellite (472 kilometres in diameter). For such a tiny body, it has a remarkably diverse surface, probably more varied even than Enceladus, although the number of superimposed impact craters shows that its last activity was probably billions of years ago. *Voyager 2* saw only half the globe. Half of that is heavily cratered, but is unusual in that most craters (the older ones) have smooth profiles, as if they have been mantled by something falling from above, and only younger craters are pristine. The other half of the imaged area comprises three sharp-edged terrain units described as coronae. Each is different, but they all contain complex ridged or tonally patterned terrain, including features identified as cryovolcanic lava flows (probably water-ammonia lava, as on Ariel), and pocked by pristine craters equivalent to those in the heavily cratered terrain.

An early theory about Miranda that each corona represents a fragment from catastrophic global break-up and re-accretion has been discounted. Most likely, the coronae are sites of cryovolcanism, of which only the waning phase has left recognizable flow-like traces. Mantling of the older craters in the terrain beyond the coronae may demonstrate explosive eruptions, spraying icy particles into space, some of which settled, snow-like, to subdue pre-existing topography. When and why this happened, we do not know. We are unlikely to find out until there is another mission to Uranus, which is not likely before mid-century.

Triton

Triton is Neptune's largest satellite (2,706 kilometres in diameter). Its outer part is icy, but it is dense enough to have a substantial rocky core. When *Voyager 2* flew past in 1989, it found polar caps of frozen nitrogen ice (previously detected spectroscopically from Earth). Like the carbon dioxide in Mars's polar caps, these probably shrink in summer, by sublimation rather than by melting, adding their content to Triton's thin but respectable atmosphere which is made largely of nitrogen. The stable 'bedrock' ice forming Triton's crust appears to be a mixture of methane, carbon dioxide, carbon dioxide, and water. There may be ammonia too, which is almost invisible to optical spectroscopy.

The best images of Triton have a resolution of about 400 metres per pixel. They reveal a geologically complex surface beyond the polar cap, including various landforms that may have been created cryovolcanically (Figure 24). Impact craters occur everywhere, but not in vast numbers, and it is possible that much of the surface is less than a billion years old. Triton is also remarkable for having geysers that erupt through the polar cap, lofting dark particles to a height of about 8 kilometres. There are a few high-altitude clouds made of nitrogen crystals, analogous to cirrus clouds in our own atmosphere.

Only the south polar cap was seen by *Voyager 2*, because most of the northern hemisphere was in darkness. Triton's seasons are peculiar, resulting from a combination of Neptune's 29.6° axial tilt

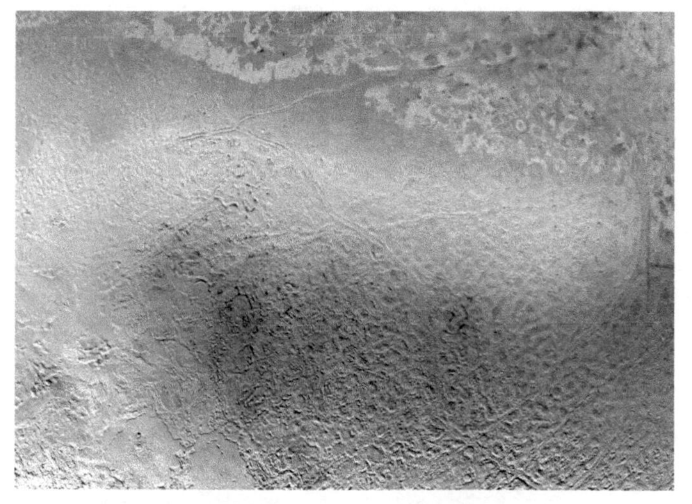

24. A mosaic of *Voyager* images covering a 2,000-kilometre-wide region of Triton. South is towards the top, and sunlight is coming from the upper right. The ragged edge of the south polar cap runs diagonally across the top of the image. Long, narrow, curved ridges (sulci) may be fissures where cryovolcanic icy magma was erupted. The smooth plains and basins in the lower left are probably expanses of cryovolcanic lava. The dimpled area in the centre and lower right is called 'canteloupe terrain', by visual analogy to the skin of a melon, but its origin is unknown

added to the 21° inclination of Triton's orbit. Further to this, Trion's orbital plane precesses about Neptune's axis so that a full seasonal cycle on Triton equates not to Neptune's 164-year orbital period, but to a 688-year cycle, with 164-year subcycles superimposed. During the full cycle, the subsolar latitude on Triton ranges between 50° north and 50° south. By chance, when *Voyager 2* made its fly-by, Triton was approaching extreme southern summer, with the Sun overhead at nearly 50° south, so a large part of the northern hemisphere was in darkness and could not be seen. The sunlit southern polar cap showed signs of being in retreat, and its sublimation to gas was verified by observations from Earth in 1997 showing that atmospheric pressure had doubled in the eight years since the *Voyager* encounter. Meanwhile, the unseen north polar cap was probably growing, as atmospheric nitrogen condensed onto the frigid surface.

Chapter 5
Asteroids

No book about planets would be complete without a discussion of asteroids, because these are the most common objects to hit planets in the inner Solar System (where asteroid impacts are about ten times more common than comet impacts). In addition, the largest asteroid, Ceres, is officially classified as a dwarf planet.

Shapes, sizes, and compositions

Ceres was the goal for NASA's *Dawn* spacecraft, which spent 43 months orbiting it between 2015 and 2018, having already spent the year beginning in July 2011 at Vesta, the second most massive asteroid. A few smaller asteroids have already been visited by spacecraft, providing images (Figure 25) that confirm their irregular shapes. Visualize a pock-marked potato scaled up to any size between tens of metres and a few hundred kilometres, and you should have a serviceable mental image of a typical asteroid, though some are different. Telescopically observed periodic variations in asteroids' brightness show that mostly they take only a few hours to rotate. Generally, rotation is at right angles to their length, so they rotate like sausages twirled on a cocktail stick.

About 1 asteroid in 50 probably has its own satellite, and it was lucky that Ida, the second asteroid to be visited by a spacecraft

25. Images of asteroids at different scales. Top: Ida, a 54-kilometre-long main belt asteroid, with its tiny satellite Dactyl to its right. Lower left: Ryugu, a 0.9-kilometre-diameter spinning top-shaped near-Earth asteroid. Lower right: Itokawa, a 0.5-kilometre-long Earth-crossing asteroid. There are many impact craters visible on Ida, but the much smaller Ryugu and Itokawa are boulder-strewn

when *Galileo* flew past in 1993, turned out to be one of these. That was the first confirmed discovery of an asteroid satellite, but subsequently many more have been found using advanced telescopic techniques, such as adaptive optics to compensate for the shimmering of the Earth's atmosphere. Asteroid satellites range from the comparatively tiny up to sizes similar to the main asteroid. In fact, the asteroid named Antiope appears to consist of two mutually orbiting bodies of indistinguishable 110-kilometre size, whose centres are only about 170 kilometres apart. So far, there are two asteroids known to possess two small satellites each. Some asteroid satellites may be fragments from a collision, and others may be captured objects. Neither case is readily explicable, because it is hard to end up with objects orbiting rather than flying apart.

Asteroid densities have been measured between 1.2 and 3.0 g/cm^3. However, stony meteorites, which are clearly bits of asteroid, have densities of about 3.5 g/cm^3 and stony-iron meteorites have densities close to 5.0 g/cm^3, so none of the measured asteroids can be an intact solid body. Rather, they must be porous rubble piles. Some, such as Itokawa, visited by the Japanese probe *Hayabusa* in 2005 (Figure 25), and others whose shape has been determined by radar, appear to be 'contact binaries' consisting of two main masses joined at a narrow waist. However, the numerous boulders on much of Itokawa's surface show that the two main masses are themselves composed of many pieces.

Asteroids are not strongly coloured, but can be grouped into several classes according to their reflectance spectrum. There are three main types. S-types have the characteristics of silicate rock, and are evidently the same material as stony meteorites. They make up the majority of asteroids with orbits between about 2.0 and 2.6 AU from the Sun, whereas from 2.6 to 3.4 AU, C-types, having the characteristics of carbonaceous chondrite meteorites, are the most common. Beyond 3.4 AU, asteroids tend to be dark and somewhat red in colour. These are dubbed D-type, and may be coloured by a tarry surface residue formed from carbonaceous material during prolonged exposure to solar radiation (space weathering). These tarry substances are usually referred to as 'tholins', a term coined from the ancient Greek word for 'mud' by the American astronomer Carl Sagan (1934–96).

Scattered here and there are asteroids that appear largely metallic (M-type), clearly related to iron meteorites, and a few that appear to have basalt on their surface, notably Vesta, from which they take their designation V-type. These, or their now fragmented parent body, may have once been hot enough for internal melting and volcanic eruptions.

Asteroid orbits

Most known asteroids (equivalent to about 4% of the Moon's mass) have orbits lying between the orbits of Mars and Jupiter, in the so-called 'asteroid belt'. Over 3,000 main belt asteroids have been documented. More than half the total mass of these resides in the four largest examples, Ceres, Vesta, Pallas, and Hygeia, with diameters of 950, 530, 540, and 430 kilometres respectively (Vesta is denser than Pallas, so is more massive despite being slightly smaller). Undiscovered objects range down in size through individual lumps of rock to dust particles. Nevertheless, the asteroid belt is virtually empty space, and you should not think of it as replete with jostling rocks. All space probes that have been sent through the asteroid belt have survived without mishap, and even have to be steered carefully to come close enough to any asteroid to study it in passing.

Jupiter's gravity has considerable influence on main belt asteroid orbits. Notably, it prevents asteroids settling into orbits whose periods would be in resonance with its own. There are virtually no asteroids whose orbital periods are simple 4:1, 3:1, 5:2, or 2:1 ratios of Jupiter's. These correspond to average distances from the Sun (orbital semi-major axes) of 2.06, 2.50, 2.82, 3.28 AU, respectively, which are known as the Kirkwood gaps, after Daniel Kirkwood, an American astronomer who discovered and explained them in 1886. Not all orbital resonances are unstable with respect to asteroid orbits, and in fact there is a small family of asteroids whose orbital periods are two-thirds that of Jupiter (a 3:2 orbital resonance).

There are a great many more asteroids with the *same* orbital period as Jupiter. There may be more than a million of these greater than 1 kilometre in size, with a combined mass about one-fifth that of the main belt. These occur only close to locations 60° ahead of, and 60° behind, Jupiter in its orbit. Those are special places where the combined gravitational force from the Sun and Jupiter allows small objects to orbit stably, and are known as the leading and trailing Lagrangian points. Asteroids in these orbits are by convention given names of heroes from the Trojan War (Greek names 60° ahead of Jupiter and Trojan names 60° behind), but are collectively referred to as 'Trojan asteroids'.

We're all doomed!

A few asteroids are known in similar 'trojan' relationship to Mars, but Earth has no trojan companions. However, there are asteroids whose orbits cross ours, known as Earth-crossing asteroids. If you are worried about collisions, this may sound alarming, but asteroid orbits tend to be inclined to the ecliptic, so they almost always pass either 'above' or 'below' us when they cross our orbit. Only a subset of Earth-crossers are regarded as Potentially Hazardous Asteroids (PHAs), being those that can pass within 0.05 AU of the Earth (a range sufficiently close that perturbations caused by various third bodies could bring about a collision) and that are larger than about 150 metres in diameter (big enough to survive passage through the atmosphere with undiminished speed). By the end of 2009, about 1,100 PHAs had been documented, plus fewer than 100 Potentially Hazardous Comets.

The closest calculated approach by a PHA is by Apophis (350 metres long) that will come very close on Friday 13 April 2029. Soon after its discovery, in 2004, its orbit was sufficiently poorly known that there was a chance (estimated at 2.7%) of a collision, but subsequently a longer series of observations showed that it will pass safely about 30,000 kilometres above the surface. It will be back again on 13 April 2036, and because we do not know *exactly* how close it will pass in 2029, we do not know exactly how much its trajectory will be affected by the Earth's gravity during that encounter. However, the chances of collision in 2036 are vanishingly small.

An asteroid that penetrates Earth's atmosphere with undiminished speed is very dangerous. On hitting the ocean, it could cause a tsunami, and if it hits land, it excavates a crater much larger than itself and devastates the surrounding area. A 2.2-million-year-old, 130-kilometre crater named Eltanin has been discovered under the floor of the Bellinghausen Sea, in the southernmost Pacific ocean, apparently caused by an asteroid several kilometres in diameter. This would have barely been slowed by the ocean, let alone the atmosphere, before striking the sea bed. According to computer models, the resulting tsunami would have devastated the coast 300 metres above sea-level in southern Chile and 60 metres above sea-level in New Zealand. The quantity of water and dust thrown into the atmosphere might even have been the trigger for climate change leading to the migration of our ancestors, *Home erectus*, out of Africa, at about this date. The most recent collision between the Earth and a 10-kilometre 'dinosaur-killer' asteroid happened 65 million years ago, creating the 200-kilometre-diameter Chicxulub crater, now buried beneath sediment in the Yucatan peninsula of Mexico. This caused a global environmental upheaval that is widely credited as the cause of a 'mass extinction event' when about 75% of species of life on Earth were wiped out.

Catastrophes of that magnitude are mercifully rare, but statistics show that asteroid impacts rank alongside volcanic eruptions, earthquakes, and extreme weather events as potential causes of death. A 1-kilometre asteroid able to devastate coasts 3,000 kilometres from the point of impact strikes the ocean on average about every 200,000 years, whereas a 200-metre asteroid with a significantly smaller tsunami danger radius might be expected about every 10,000 years.

To categorize the hazard posed by each PHA, astronomers use a numerical system called the Torino Scale (agreed at a meeting in Turin, hence the name). This combines the energy that would be delivered and the probability of collision into a single number from 0 to 10, where 0 means negligible chance of collision and/or too small to penetrate the atmosphere, and 10 is certain impact by a ‘dinosaur-killer’ causing global catastrophe. Most PHAs exceeding 150 metres in diameter are ranked either 0 or 1 when they are discovered, and the 1s are usually downgraded to 0 when their orbit has been more adequately determined. Apophis holds the record for having temporarily held a Torino rating as high as 4 (‘Close encounter, meriting attention by astronomers; >1% or greater chance of collision capable of regional devastation’), but was downgraded to 0 in 2006.

A semi-formal linking of observatories known as Spaceguard has assumed the task of locating and categorizing all PHAs. This is important because, unlike most sorts of natural disaster when all we can do is mitigate the effects, it would be possible to prevent a collision by a PHA. To achieve this, it is necessary to change either the PHA’s speed or its direction of travel. The longer in advance this is done, the smaller the required change. There are various ways to do it, ranging from the brute-force method of equipping the PHA with a rocket motor, to the more subtle ploy of coating of one side in a reflective substance so that solar radiation-pressure does the job for you. Using a nuclear bomb to blast apart an incoming PHA is not a smart idea, because unless you could guarantee that all the fragments would be too small to penetrate the atmosphere, you might make the problem worse by causing multiple impacts.

Asteroid mining

There is a silver lining, in that asteroids could be valuable sources of raw materials. A 1-kilometre M-type asteroid contains more nickel and iron than the world’s annual consumption, and the most massive example, Psyche, the goal of a NASA mission of the same name to be launched in 2022, has enough to last for millions of years. Asteroids, especially M-types, also contain precious metals like platinum.

The investment to begin mining the first asteroid would be very great, but the potential returns are immense too. It remains to be seen whether the main value of asteroids turns out to be supply of raw materials to Earth or to space-based industries. Some near-Earth objects are probably defunct comets with remnant water-ice surviving beneath their dusty surfaces, which could be valuable as propellant and radiation shielding, as well as for drinking.

Names and provisional designations

By 1891, 332 asteroids had been discovered visually, but photography had boosted the tally to 464 within 10 years. There are now over 100,000 known objects of all types, each of which needs to be identified in some way. The IAU oversees a system of provisional designations given to each new discovery. This consists of the year of discovery plus a two-letter code coupled with numerical subscripts, corresponding to the date and sequence of discovery. The first letter (A–Y, omitting I) specifies which half-month the discovery was made in (A for January 1–15, B for January 16–31, and so on, up to Y for 16–31 December), the second letter (A–Z, omitting I, which gives 25 options) is awarded to each discovery in sequence, and a numbered subscript allows the cycle of 25 to be repeated as many times as necessary. Thus, 2011 BA would be the first body discovered in the period January 16–31 2011; 2011 BB would be the 2nd; 2011 BA_1 would be the 26th, and so on. When an object's orbit has been well determined (which may take several years), it can be awarded a permanent name, which replaces the provisional designation. For example, Apophis originally had the provisional designation 2004 MN_4 (signifying the 113th discovery during 16–30 June 2004).

The privilege of suggesting a permanent name is given to the discovery team, though some automated surveys reveal so many new objects that managers are glad of suggestions. Permanent names are a name preceded by a number, added in sequence as each new name is added. So formally we have (1) Ceres, (4) Vesta, (99942) Apophis, and so on. Available mythological names are too few for all these objects, and pretty much anything is allowed, except that names must be inoffensive and unconnected with recent political or military activity. I know several astronomers who have had asteroids named after them (by colleagues; you can't name one after yourself), and there is one called (5460) Tsenaat'a'i, which means 'flying rock' in the Navaho language. The only asteroid that I have had a hand in naming is (57424) Caelumnoctu, named in 2007 to commemorate the 50th anniversary of BBC television's long-running programme *The Sky at Night*, which in Latin is *Caelum Noctu*. We picked it from a list because its number reflects the date of the first broadcast, which was 1957 April 24 (57/4/24).

Chapter 6
Trans-Neptunian objects (TNOs

A relatively sparse population of asteroids, known as Centaurs, exists between Jupiter and Neptune. Some are dark and red, similar to the tarry (tholin-covered) D-type asteroids, but others are bluer, suggesting that much of their surfaces may be freshly exposed ice. Their orbits cross or come close to the giant planets. Such orbits are not stable, persisting for no longer than about ten million years. Probably, Centaurs are TNOs that have been scattered inwards, perhaps by a close encounter with Neptune. Further interactions with giant planets probably nudge them inwards until they become short-period comets, spending perihelion in the inner Solar System, where they are heated by the Sun and lose their volatiles in sometimes spectacular tails.

Nineteen trojan objects have been discovered close to Neptune's leading Lagrangian point, and three near the trailing point. Dynamical arguments suggest that vast numbers await discovery and that Neptune trojans may be ten times more numerous than Jupiter's.

Beyond Neptune, we reach the Kuiper belt and all the other TNOs. One family of Kuiper belt objects travel in 3:2 orbital resonance with Neptune. Members of this class, which includes Pluto, are known informally as 'plutinos', not to be confused with Plutoid, which is the official IAU term for any TNO large enough to be ranked as a dwarf planet. Plutoids can be plutinos, classical

Kuiper belt objects (lacking orbital resonance with Neptune), or Scattered Disk objects beyond the main belt. Classical Kuiper belt objects are known alternatively as 'cubewanos' ('QB_1-os') because the first Kuiper belt object to be discovered after Pluto bore the provisional designation 1992 QB_1.

Pluto and Charon

The properties of most TNOs are poorly known. However, Pluto and its satellite Charon are sufficiently large and nearby to have been studied telescopically for several decades, and parts of their surfaces were imaged in great detail when NASA's New Horizons probe flew past in 2015. Much of Pluto's surface is ancient impact craters and rugged mountains, demonstrating a strong crust of water-ice, but frozen nitrogen, methane, and carbon dioxide have been detected spectroscopically too. A 1000 km wide basin known as Sputnik Planitia, probably the scar of a very large impact, is surfaced by bright nitrogen-ice that may have oozed out of the interior, . The surface of Sputnik Planitia is divided into cells a few tens of km across that is probably a pattern resulting from convective flow of the underlying ice, and the surface lacks any obvious impact craters attesting to its young age. Around the edge of the basin, the nitrogen-ice can be seen spilling outwards via glacier-like flow into valleys in the surrounding mountains. There are dark patches on the mountainous cratered terrain that are probably tholin-rich residues that have settled out of the atmosphere. Pluto's density suggests that rock must be about 70% of its total mass, and most likely it is internally differentiated with a rocky core (and feasibly an iron-rich inner core) overlain by a mantle made mostly of water-ice (possibly including a liquid internal ocean) topped by a more volatile-rich crust.

Near perihelion (which happened most recently in 1989), Pluto has a nitrogen-rich atmosphere possibly denser than Triton's. Because Pluto's gravity is so weak, an imaginary shell enclosing 99% of its atmosphere would extend to 300 kilometres above the surface, whereas for the Earth the equivalent height is only 40 kilometres. Much of Pluto's atmosphere is expected to condense onto the surface while distance from the Sun increases from 4.5 billion kilometres at perihelion to 7.4 billion kilometres at aphelion in 2113. At the time of *New Horizon's* flyby the surface pressure was about 100,000 times less than at sea-level on Earth but the atmosphere was still substantial enough to contain numerous haze layers up to an altitude of 200 km.

Pluto's 6.4-day rotation period is the same as the orbital period of its largest satellite, Charon, which also rotates synchronously. This relationship is a result of strong tides, and means that Pluto and Charon permanently keep the same faces towards each other. The pair are more evenly matched in size and mass than any other planet or dwarf planet and its own largest satellite. Charon's mass is about 12% of Pluto's, and it orbits at a distance of only about 17 Pluto radii from Pluto's centre. For comparison, the Moon's mass is only 1.2% of Earth's and its orbital radius is 60 Earth radii. Charon's proximity to Pluto explains why it remained undetected until 1978. Pluto has four much smaller satellites, (Table 7) that orbit in Pluto's equatorial plane close to 3:1, 4:1, 5:1, and 6:1 resonance with Charon.

Seen from Pluto's surface, Charon would look eight times wider than the Moon does from Earth. Because their relative masses are so similar, their common centre of mass (their 'barycentre') is not inside Pluto but at a point in space between the two bodies. Although double asteroids such as (90) Antiope and double Kuiper belt objects such as 2001 QW_{332} (200-kilometre diameter twins) are known, Pluto-Charon is the most evenly matched pair among bodies large enough to count as planets or dwarf planets.

Charon's density is less than Pluto, but still enough for a substantial rocky core. Its surface is dominated by water-ice with traces of ammonia, and around the north pole (the only pole visible during the flyby) is a patch discoloured by tholins presumed to have drifted across from Pluto's atmosphere.

Charon is a darker, drabber world than Pluto. The north is more heavily cratered than the south, which tells us that the terrain in the north is older, whereas the south has been subjected to resurfacing. How this may have occurred is hinted at by a rift named Serenity Chasma that stretches across the Pluto-facing hemisphere near the equator, where the most ancient surface has been ripped apart and then flooded cryovolcanically in a manner reminiscent of Uranus's moon Ariel.

The rest

Table 7 lists Pluto and the nine largest other TNOs as ranked in 2020. Of these, Eris, Makemake, and Haumea are officially recognized as dwarf planets. Eris, although marginally smaller than Pluto is denser and so contains more mass. Haumea is flattened, either because of its rapid rotation (less than 4 hours) or resulting from collision. These are classical Kuiper belt objects, except Eris and Gonggong (Scattered Disk objects), Ixion

Table 7 The largest trans-Neptunian objects

Name	Diameter /km	Mean distance from Sun/AU	Orbital period /years	Known satellites
Pluto	2377	39.4	248.0	Charon (1,208 km), Styx (16 km), Nix (50 km), Kerberos (19 km), Hydra (65 km)
Eris	2326	67.7	557	Dysnomia (<250 km)
Makemake	1,400–1,450	45.8	309.9	Unnamed (170 km)
Haumea	1,400	43.1	283.2	Hi'iaka (310 km), Namaka (170 km)
Quaoar	1,100	43.6	288.0	Weywot 170 km
Gonggong	1,180–1,280	67.3	552.5	Xiangliu (100 km)
Sedna	910–1,080	525.9	12,059	None
Orcus	890–940	39.2	245.3	Vanth (450 km)
Varuna	650–800	43.1	283.2	(1 suspected)
Ixion	600–640	39.7	248.9	None

(plutino), and Sedna, which is an oddity way beyond the Scattered Disk, in a highly elliptical orbit with aphelion at 975 AU.

Apart from Pluto, the sizes of these objects are poorly known (even those for which a single round number is given in the table). Their dimensions are usually estimates based on assumptions about their albedo (the percentage of the incident sunlight they reflect). If they are less reflective than assumed, they must be larger, but if they are more reflective, then they must be smaller. Size estimates can be improved by measuring thermal radiation from their surfaces, but they are so cold (−230 °C or less) that this can be achieved only by telescopes in space, above the Earth's atmosphere.

TNOs range in colour from red (probably widespread tholins across their surfaces) to blue-grey (exposed ice or amorphous carbon). Haumea is one of the blue-grey ones, and its mass (derived from the orbits of its satellites) shows that its density is greater than Pluto's, so it must have a relatively high non-ice content. On Quaoar, crystalline ice and ammonia hydrate have been detected spectroscopically, suggesting recent resurfacing. This would require either geological activity or a major impact event to generate ejecta sufficiently widespread to dominate the spectrum.

Between 2% and 3% of TNOs are known to have satellites, which is similar to the abundance of asteroids with satellites. The proportion is higher among the larger TNOs and poses problems in trying to account for their origin.

In 2019 *New Horizons* passed within 3500 km of a small TNO. formerly catalogued as 2014 MU69 but renamed Arrokoth, and revealed it to consist of two lobes (20 km and 18 km in diameter) that are presumed to have joined in a low-speed collision. At nearly 44 AU from the Sun during the fly by, this is the most distant object yet visited by a spacecraft. If *New Horizons* remains healthy it is hoped that it will have a close encounter with an even more distant TNO sometime in the 2020s.

A trans-Neptunian planet?

Most astronomers accept that we have discovered all the large objects belonging to our Solar System. Certainly, there can be nothing of planetary size hiding in the Kuiper belt. If such an object were present, then the Kuiper belt would be unstable. However, there remain two possibilities for an outlying planet (popularly dubbed 'Planet X') that have not yet quite gone away. One is that there is a 5–10 Earth-mass object in an inclined and eccentric orbit beyond 300 AU from the Sun. The presence of such a large body (perhaps originally ejected outwards by a close encounter with Neptune) might help to explain an observed sudden drop-off in the population of the Kuiper belt beyond 48 AU, known as the 'Kuiper cliff'. It might also account for the extreme scattering evidenced by objects such as Sedna.

The second possibility arises because long-period comets originate preferentially from a particular region of the sky, rather than coming in from random directions. It has been suggested that these were dislodged from the Oort Cloud by a Jupiter-mass body about 32,000 AU from the Sun. This would be hard, but not impossible, to detect by telescope. A 'planet' so far out need not be gravitationally bound to the Sun, but could be just a chance wanderer through interstellar space, possibly escaped from the planetary system of another star.

Chapter 7
Exoplanets

There is no longer any doubt that planets are common around other stars. Until comparatively recently, this was a matter for speculation, but by 2021 over 3000 stars had been proven to have at least one planet orbiting them. Allowing for how difficult it is to make these detections, it is clear that the majority of Sun-like stars must be accompanied by planets. So too are lower-mass, fainter stars known as red dwarfs. To avoid confusion, professionals usually refer to them as 'extrasolar planets' or 'exoplanets'. The exoplanet tally excludes exotic dim objects exceeding 13 Jupiter-masses, which is the threshold above which nuclear fusion of deuterium (heavy hydrogen) can occur. Those are called 'brown dwarfs' and are regarded as more star-like than planet-like.

Detection methods

Evidence that most young Sun-like stars have a surrounding ring of dust began to accumulate in the late 1970s. Initial clues came from the influence of dust on a candidate star's infrared spectrum, then images of dust discs began to be obtained in the 1980s. Irrespective of whether these discs are like the solar nebula before planets formed or are remnant dust surviving in the equivalent of a star's Kuiper belt, their very existence showed that there ought to be plenty of planets out there too. The first definite exoplanet discovery was made in 1995, after which discoveries gathered pace year by year.

Radial velocity

The first discovery, and the majority during the first decade and a half, was made by detecting slight changes in a star's radial velocity. This is the speed at which a star is travelling towards or away from the Earth, irrespective of any movement across the line of sight. Radial velocity changes can be determined to a remarkable precision of a fraction of a metre per second by measuring shifts in the wavelength at which characteristic absorption lines occur in a star's spectrum. These shift to shorter wavelengths ('blue shift') if the star is moving towards us and to longer wavelengths ('red shift') if the star is travelling away, in a phenomenon called Doppler shift. Variations in radial velocity had long been used to measure orbital speeds (and hence to infer masses) of double stars, but the tiny influence of a much less massive exoplanet on a relatively much more massive star requires very sensitive modern instrumentation. Radial velocity changes caused by the Earth's own orbital motion have to be accounted for before the more subtle changes attributable to the tug of the exoplanet on its star become apparent.

The gravitational attraction between star and exoplanet depends on the sum of their masses. Fortunately, for Sun-like stars there is a well-understood relationship between stellar spectral type and mass. Knowing this, we can use the period and magnitude of radial velocity changes to determine the mass of the exoplanet responsible for the forward and back motion of the star. Usually, there is no independent measure of the orientation of an exoplanet's orbital plane, and unless the orbital plane is exactly edge-on to our line of sight, the true change in velocity must be greater than what we detect. However, statistical arguments (based on assuming randomly oriented orbital planes) show that the majority of masses can be no more than twice the estimate based on assuming that the orbit is edge-on to us.

The radial velocity method works best for large planets orbiting close to their star, because large mass and close proximity both lead to the greatest changes in the star's radial velocity. Thus the first exoplanets to be detected tended to be more massive than Jupiter but orbiting only a fraction of an AU from their star.

Discovery of these so-called 'hot Jupiters' caused quite a stir, because they are well inside their stars' ice line and cannot have formed where we now see them. It is now accepted that they grew further out and then migrated inwards, and this has reopened the debate about the extent of planetary migration in our own Solar System's early history. If Jupiter's migration had continued inwards, it would have either destroyed or scattered each terrestrial planet in turn. Improved and additional techniques for exoplanet detection have begun to find rocky planets, showing that the preponderance of 'hot Jupiters' in the early discoveries was merely a selection effect resulting from ease of discovery. Earth-mass planets have begun to be detected in numbers, and in 2016, the radial velocity method discovered a 1.2 Earth-mass exoplanet orbitting our nearest star, the red dwarf Proxima Centauri only 4.2 light years away.

Transits

The most prolific method for discovering exoplanets, which has found about 2/3 of the total known in 2020, has been to search for 'transits', when a tiny fraction of a star's light is cut off during passage of an exoplanet in front of it. Most transits are discovered by repeated scans of likely stars using automated telescopes, originally from the ground more productively by dedicated telescopes in space, notably by NASA's aptly-named Kepler telescope (2009-2018) that found more than 2600 exoplanets.

A transit can happen only if the exoplanet's orbital plane lies close to our line of sight, which statistically should apply to only about half a percent of all exoplanetary systems. The dimming of the starlight is slight, but is greatest for the largest exoplanets and occurs more often (and so is more likely to be detected) for exoplanets orbiting close to their star. The exact amount by which the starlight dims can be used to deduce the size of the planet compared to its star. The duration of the transit gives us clues to orbital speed and orbital radius, but follow-up radial velocity measurements can better characterize the system. Because a transit demonstrates that the orbital plane lies in our line of sight, masses derived by the radial velocity method are true values rather than minimum estimates. One of the smallest exoplanets now known was characterized in this way; it is a Mars-sized, 0.07 Earth-mass, body that is the smallest of three orbitting a red dwarf star known as Kepler-138.

Imaging and other methods

Exoplanets are exceedingly challenging to image, because they are so much fainter than their stars. Exoplanets have been imaged at only about a hundred stars. As you might expect, these are all Jupiter-sized or larger, mostly orbiting at tens or even hundreds of AU. In 2008, an adaptive optics image obtained from infrared telescopes in Hawaii showed three exoplanets orbiting a young Sun-like star (catalogued as HR 8799) at 24, 38, and 68 AU, and a fourth was found at 15 AU in 2010. Beyond them is a dust disc at about 75 AU.

Another method for exoplanet detection, called 'astrometry', is based on very precise measurement of a star's position in the sky. Any unseen orbiting companion will tug the star from side to side. Astrometry seeks to detect this, instead of radial velocity changes along the line of sight. The motion is greatest if caused by a massive planet in a large orbit, in contrast to methods more sensitive to small orbits. The first confirmed success of the as trometric method was in 2002, when the Hubble Space Telescope documented sideways wobble of the star catalogued as Gliese 876, refining what we knew about a 2.6 Jupiter-mass planet orbiting at 0.20 AU already detected by radial velocity changes.

A wholly different technique takes advantage of random (and never to be repeated) exact alignment between a foreground star and a background star. The foreground star acts as a 'gravitational microlens' that amplifies the light from the background star. The detected brightness of the background star rises and then falls over a duration of several weeks. If the foreground star has a fortuitously placed exoplanet, this will cause a brief spike in the brightness (lasting a few hours or days) superimposed on the slower rise and fall. By late 2020, microlensing had discovered a about 130 exoplanets.

Naming exoplanets

Exoplanets are catalogued by adding a lower-case letter after the name or designation of their star. The first to be discovered is b, the second is c, and so on (a is not used). Thus, the first exoplanet of Gliese 876 is Gliese 876 b, and two subsequently discovered exoplanets in the same system are Gliese 876 c and Gliese 876 d. This convention is messy, because the letters bear no relationship to the order of exoplanets within multiple systems. The IAU invited the public to vote for names to be given to a small subset of exoplanets in 2013, but this was not intended to replace the formal catalogue designation. Perhaps it is wise for us not to impose names. Maybe the natives already have perfectly good names for their homes.

Multiple exoplanet systems

Multiple exoplanets were known at more than 700 stars by 2021. Sometimes a combination of detection techniques provides this information, but radial velocity or precise timing of serial transits alone can do the job: it is just a matter of unravelling progressively more subtle periodic variations.
So far as we can tell, our Solar System is not exactly typical, but it is not all that unusual either. Exoplanet types that we lack here, but which have been found in abundance elsewhere, are 'super-Earths' (diameter 1.2-2, and mass 1.4-10 times Earth's) and 'mini-neptunes'. At the right temperature, a super-Earth is likely to be a 'waterworld', with a global ocean accounting for about 10% of its mass.

The Sun-like star with the most currently known exoplanets is Kepler-90. This is a slightly hotter star than the Sun, with eight exoplanets known from analysis of transits. The six innermost are super-Earths and the outer two are Jupiter-like.

Radial velocity studies have shown seven confirmed (plus two unconfirmed) exoplanets orbitting a similar star, HD 10180. Of its confirmed exoplanets, the closest is an Earth-mass world only 0.02 AU away. The others are similar in mass to Neptune or Saturn, and none is further from the star than 3.5 AU, so it is a very compact system.

Among red dwarfs, there is a known family of seven exoplanets orbitting the star TRAPPIST-1, discovered by transit techniques. Four of them are about Earth-mass and the others are less massive. The most distant is only 0.06 AU from the star, but because red dwarfs are so faint their habitable zones are also close. In this case, the 3rd and 4th exoplanets out are calculated have the right temperature to in the habitable zone.

Study

We have little direct information about any exoplanet. If we determine mass (by radial velocity or astrometry), we can infer size by assuming a likely density. A transit will reveal size, which can also be deduced by imaging (based on brightness and assumed albedo). From size, we can deduce mass if we assume a density. Distance from its star gives us a fair idea of surface (or atmospheric) temperature, but this depends also on albedo and the mixture of greenhouse gases in any atmosphere, so there is a conside rable margin for error.

The next major advance in the study of exoplanets will probably come as we improve our ability to analyse their atmospheric composition. So far this has been achieved only for a few, mostly large, exoplanets in close orbit. To do better will probably require telescopes in space, capable of isolating and analysing the visible and infrared spectra of individual exoplanets – most importantly terrestrial ones. Several abundant atmospheric gas species can be identified by their characteristic absorptions. Detection of a pair of gases that ought not to co-exist under conditions of simple chemistry, such as oxygen and methane, may be the first evidence we obtain of life affecting an exoplanet's atmosphere in the same way that the Earth's atmosphere has been radically changed.

Life on exoplanets

There are about 10,000 million Sun-like stars in our galaxy (about 1 in 10 of the total stars). Exoplanets must be abundant. They have been found orbiting about half of the adequately studied Sun-like stars, and at red dwarfs too, which are even more abundant. Most discoveries so far have been giant planets, because those are the easiest to find. Although terrestrial planets are unlikely to have survived the inward migration of a 'hot Jupiter', plenty of Earth-mass planets have by now been located.

The question of how many exoplanets might host life is a vexing one. Let's take a very conservative estimate that on average only 1% of Sun-like stars have a terrestrial planet orbiting in a long-duration habitable zone. That would give 100 million habitable terrestrial planets in our galaxy, and there are probably at least as many habitable satellites orbiting giant exoplanets. When we consider red dwarfs, any exoplanets close enough to be in the circumstellar habitable zone must almost certainly be tidally-locked in step with their orbital period (like most satellites in our Solar System). Such exoplanets would have (to us) a curious climate, with a permanent day side and a permanent night side, but any atmosphere would help to ameliorate temperature extremes and they would not necessarily be uninhabitable as was once assumed. It is not unreasonable to suppose that there are as many habitable exoplanets orbiting red dwarfs as there are orbitting Sun-like stars.

The next step in the chain of logic is far less certain. Given the conditions required for life, how likely is it that life will begin? The building blocks for life are not a limiting factor. We know that space is awash with organic molecules, and that water is abundant too, so most exoplanets in a habitable zone will have all the necessary ingredients for carbon-based life. That means 'life as we know it' (as they say on *Star Trek*) without delving into speculation about other forms of life reliant on exotic chemistries.

The ease or difficulty with which life can spontaneously arise is a big gap in our understanding. Many (myself included) hold that with countless trillions of appropriate organic molecules in an exoplanet's ocean, and with millions of years to play with, life will inevitably start. Once life has spread, it is hard to see how it can be completely eradicated, but if it was, it could presumably restart just as readily.

We know that it took life on Earth less than 500 million years to establish a permanent footing. The abundance of life in the galaxy (and, by implication, in the cosmos beyond) will remain unproven until we detect signs of life on exoplanets. Even if we were to find current (or past) life on Mars, Europa, or Enceladus, we ought not to leap to the conclusion that life had begun there independently, because objects in the Solar System are not totally isolated from each other. Microbes could survive transport from one to another inside fragments of impact ejecta. Europan life could have come from Earth, and it is conceivable that life on Earth arrived on a meteorite from Mars.

Is anybody out there?

If there is life around other stars, what about intelligent life? Let's speculate rationally. So far as we know, biological intelligence requires multicellular life. If microbial life begins, how likely is it that subsequent evolution leads to multicellular organisms? You can take your pick on this issue. It took a couple of billion years to happen on Earth.

After multicellular life appears, will competition for survival drive Darwinian evolution as on Earth? Intelligence is one factor that confers an advantage, so how inevitable is intelligence?

Even on a conservative figure of 100 million habitable terrestrial planets in the galaxy, plus a pessimistic view that life has only a 1 in 100 chance of starting, that still leaves a million worlds with life, of which the Earth is one. It would be strange (and awe-inspiring) if Earth were the only or first planet out of all that number ever to host intelligence. But if life is so abundant, and if intelligence is a common outcome of life, then where is everybody? Unless it arises exceedingly rarely, or does not last long (for example, our own civilization could succumb to wars, various natural disasters, or self-made climate change), then the galaxy ought to be teeming with intelligence.

Intelligent life would not have to be indigenous to where we find it. Although the distances between stars are vast, it is perfectly feasible to travel between them. You do not need faster-than-light travel – all you need is determination and patience. Imagine a spaceship big enough to house hundreds of people, which would take 100 years to travel to a habitable exoplanet of a star 10 light years away. We could build such a ship ourselves, using technology foreseeable in the next few decades. One or two generations of the crew would live and die en route (unless some kind of suspended animation is used), and it would be very much a one-way trip. If we were to send such colonists to all the nearby habitable exoplanets it would not be long before successful colonies had the capacity to launch their own colony ships, and so on. The galaxy is 100,000 light years across. Even if a wave of colonization takes 1,000 years to spread 10 light years, the entire galaxy could be colonized in only 10 million years. (either by living organisms or self-replicating robots). Catastrophes wiping out whole worlds or failures of individual colony ships would be insufficient to derail the process once it was underway.

The galaxy is more than 10 *billion* years old. If intelligent life is abundant, there has already been ample time for countless previous species to have colonized the galaxy. This is the Fermi Paradox, named after comments made by the American physicist Enrico Fermi in 1950. Extraterrestrial civilizations ought to be numerous, but there is no sign of them: no artificial signals detected from space (despite scans of the sky by teams working under the banner of SETI – Search for Extra-Terrestrial Intelligence), no sign of great works of astronomical engineering, and no credibly documented alien visitors to Earth. Is intelligent life rare, after all, or are we too stupid to see the evidence? One day, I hope we will find out.

Further reading

There is a rich literature associated with astronomy and planetary science. The trouble is that, the longer or more specialized the book, the faster it goes out of date. On the other hand, some (not all!) websites are frequently updated. To help you discover more about planets, I suggest a few of the best books and several appropriate entry points to the internet.

General

J. K. Beatty, C. C. Peterson, and A. Chaikin (eds.), *The New Solar System*, 4th edn. (Sky Publishing Corporation and Cambridge University Press, 1999). This covers the lot. Each chapter is written by a specialist author. Badly dated in parts, but it remains a highly accessible classic.

D. A. Rothery, I. Gilmour and M. A. Sephton (eds.), *An Introduction to Astrobiology* (Cambridge University Press, 2018, 3rd edn.). The second of two volumes based around an Open University course on planetary science, written at early undergraduate level. This one covers life, Mars, Europa, Enceladus and Titan as potential habitats, and considers the characteristics of exoplanets.

D. A. Rothery, N. McBride and I. Gilmour (eds.), *An Introduction to the Solar System* (Cambridge University Press, 2018, 3rd edn.). The first of two volumes based around an Open University course on planetary science, written at early undergraduate level. It covers all the major components of the Solar System, except the Sun.

S. A. Stern (ed.), *Our Worlds: The Magnetism and Thrill of Planetary Exploration* (Cambridge University Press, 1999). Easy but informative reading. Each chapter is a personal account by one of the leading practitioners.

D. A. Weintraub, *Is Pluto a Planet?* (Princeton University Press, 2007). If you've read this far, then you already know the answer to the question posed by this book's title. However, it covers much more than that, being an historical account of human perception of planets from ancient times right up to the recent squabbles over the classification of TNOs.

Terrestrial planets

M. Hanlon, *The Real Mars* (Constable, 2004). A science writer's perspective on Mars, simply written and beautifully illustrated.

J. S. Kargel, *Mars: A Warmer Wetter Planet* (Springer Praxis, 2004). One leading scientist's personal view of the role of hidden water on Mars.

R. M. C. Lopes and T. K. P. Gregg (eds.), *Volcanic Worlds: Exploring the Solar System's Volcanoes* (Springer Praxis, 2004). A popular account, with chapters by specialist authors dealing with volcanism on each terrestrial planet, the Moon, Io, and icy satellites.

R. G. Strom and A. L. Sprague, *Exploring Mercury* (Springer Praxis, 2003). This is the best review of Mercury that I know, but written before MESSENGER began to study the planet.

Asteroids

J. Bell and J. Mitton (eds.), *Asteroid Rendezvous: NEAR Shoemaker's Adventures at Eros* (Cambridge University Press, 2002). A well-illustrated and popular account of the findings of the first probe to orbit and then crash onto an asteroid.

Giant planets

F. Bagenal, T. Dowling, and W. McKinnon (eds.), *Jupiter: The Planet, Satellites and Magnetosphere* (Cambridge University Press, 2004). A fat volume with 26 chapters written by specialist authors. Will take you much further than the current book.

E. D. Miner and R. R. Wessen, *Neptune: The Planet, Rings and Satellites* (Springer Praxis, 2002). A much slimmer and more simply written volume. Unlikely to date badly.

Satellites

D. A. Rothery, *Moons: A Very Short Introduction* (Oxford University Press, 2015). A companion book to this one, focussing on natural satellites.

R. Greenberg, *Unmasking Europa* (Springer, 2007). A clear and authoritative account of Europa, including some scathing passages about how Greenberg's research team had to struggle against the establishment to gain acceptance for their thin ice interpretation.

R. Lorenz and J. Mitton, *Titan Unveiled* (Princeton University Press, 2008). The first author is a key member of the *Cassini-Huygens* team that explored Titan, so this is an insightful account. However, it was written before Titan's lakes were fully recognized.

D. A. Rothery, *Satellites of the Outer Planets*, 2nd edn. (Oxford Unversity Press, 1999). Written by myself, this is an account of large satellites from Jupiter to Neptune at a level that should suit if the current book has left you wanting more. It includes some Galileo findings, but predates the *Cassini-Huygens* mission to Saturn so is out of date in parts.

Exoplanets

H. Klahr and W. Brander (eds.), *Planet Formation* (Cambridge University Press, 2006). More technical than most others in this list, this volume is based on papers presented at a conference in 2004. It looks at planet formation in the light of modern theories for our Solar System and discoveries of exoplanet systems.

F. Casoli and T. Encrenaz, *The New Worlds: Extrasolar Planets* (Springer Praxis, 2007). The most up-to-date popular account of exoplanets that I could find.

Websites

The following websites were accessed 24 March 2021.

General

www.nasa.gov

NASA's home page. Click on the links here for news about missions or individual Solar System bodies.

Images

photojournal.jpl.nasa.gov/
An archive of NASA images of Solar System bodies.

http://www.esa.int/esa-mmg/mmghome.pl
Multimedia gallery provided by the European Space Agency.

http://www.isas.ac.jp/e/index.shtml
Japan's Institute of Space and Astronautical Science (ISAS), with links to images and movies from Japanese missions.

arc.iki.rssi.ru/eng/index.htm
The Russian Space Research Institute (IKI). Follow the link to Planetary Exploration for access to images and information from Russian (and former Soviet) missions.

hubblesite.org/gallery/
Gallery of images from the Hubble Space Telescope, searchable by name of planet.

Maps and nomenclature

https://www.usgs.gov/centers/astrogeology-science-center/maps
A site run by the Astrogeology Science Center of the United States Geological Survey where you can view geologic and topographic maps of planets and other Solar System bodies.

planetarynames.wr.usgs.gov/
A gazetteer of nomenclature on planets, satellites, and asteroids. Hosted by the United States Geological Survey, Astrogeology Research Program on behalf of the International Astronomical Union (IAU). Contains all you need to know about naming conventions, and up-to-date searchable lists of names of all kinds of features on each body.

News and data

http://nssdc.gsfc.nasa.gov/planetary/
Has links for each planet and other classes of body, taking you to fact sheets and much more.

http://www.minorplanetcenter.org/iau/mpc.html
Website of the IAU Minor Planet Center (at the Smithsonian Astrophysical Observatory). Especially good information on near-Earth objects.

www.boulder.swri.edu/ekonews/
Electronic newsletter about the Kuiper belt, plus various useful links.

www.exoplanet.eu
The Extrasolar Planets Encyclopedia. Includes a frequently updated catalogue tracking the current tally of known objects, and also tutorials on the various methods of detecting exoplanets.

http://www.planetary.org/home/
The Planetary Society. An international (US-based) society promoting planetary exploration. A good source of relevant news and comment.

Index

COSMOLOGY

A Very Short Introduction

Peter Coles

What happened in the Big Bang? How did galaxies form? Is the universe accelerating? What is 'dark matter'? What caused the ripples in the cosmic microwave background?

These are just some of the questions today's cosmologists are trying to answer. This book is an accesible and non-technical introduction to the history of cosmology and the latest developments in the field. It is the ideal starting point for anyone curious about the universe and how it began.

'A delightful and accesible introduction to modern cosmology'

Professor J. Silk, Oxford University

'a fast track through the history of our endlessly fascinating Universe, from then to now'

J. D. Barrow, Cambridge University

www.oup.co.uk/isbn/0-19-285416-X

PARTICLE PHYSICS

A Very Short Introduction

Frank Close

In this compelling introduction to the fundamental particles that make up the universe, Frank Close takes us on a journey into the atom to examine known particles such as quarks, electrons, and the ghostly neutrino. Along the way he provides fascinating insights into how discoveries in particle physics have actually been made, and discusses how our picture of the world has been radically revised in the light of these developments. He concludes by looking ahead to new ideas about the mystery of antimatter, the number of dimensions that there might be in the universe, and to what the next 50 years of research might reveal.

http://www.oup.co.uk/isbn/0–19–280434–0

ONLINE CATALOGUE
A Very Short Introduction

Our online catalogue is designed to make it easy to find your ideal Very Short Introduction. View the entire collection by subject area, watch author videos, read sample chapters, and download reading guides.